An OPUS book

Science and Technology in World Development

Robin Clarke

Science and Technology in World Development

Foreword by
AMADOU-MAHTAR M'BOW,
DIRECTOR-GENERAL OF UNESCO

Oxford   New York
OXFORD UNIVERSITY PRESS/UNESCO
1985

Oxford University Press, Walton Street, Oxford OX2 6DP

London New York Toronto
Delhi Bombay Calcutta Madras Karachi
Kuala Lumpur Singapore Hong Kong Tokyo
Nairobi Dar es Salaam Cape Town
Melbourne Auckland

and associated companies in
Beirut Berlin Ibadan Mexico City Nicosia

Oxford is a trade mark of Oxford University Press

Published by arrangement with Unesco and sold
throughout the world by Oxford University Press

First published 1985 by Oxford University Press and the
United Nations Educational, Scientific, and Cultural Organization,
7, Place de Fontenoy, 75700 Paris
as an Oxford University Press paperback and simultaneously in a hardback edition

British Library Cataloguing in Publication Data

Clarke, Robin
Science and technology in world
development.—(OPUS)
1. Science—Social aspects   2. Technology—Social aspects
I. Title   II. Series
303.4'83   Q175.5
ISBN 0-19-219195-0
ISBN 0-19-289176-6 Pbk

Library of Congress Cataloging in Publication Data
Clarke, Robin.
Science and technology in world development.
(An OPUS book)
Includes bibliographical references and index.
1. Science—Social aspects.   2. Technology–Social aspects.
I. Title   II. Series: OPUS.
Q175.5.C53 1985   303.4'83   84-7169
ISBN 0-19-219195-0
ISBN 0-19-289176-6 (pbk.)

Set by Hope Services, Abingdon
Printed in Great Britain by
Richard Clay (The Chaucer Press) Ltd.
Bungay, Suffolk

Foreword by the Director-General of Unesco

One of the fundamental requirements of our times is that the developing countries should have access to the vast possibilities that science and technology offer for their progress and the well-being of their peoples.

The efforts so far made in these countries to develop research and training in the various fields of science and technology have in most cases been stimulated by international scientific co-operation.

Such co-operation can indeed be of great help in mobilizing national resources and organizing them for purposes of development. At the same time, by facilitating exchanges of ideas and contacts between individuals, by promoting the dissemination of the findings of research carried out in various establishments, and by encouraging the co-ordinated study of common problems at regional level, co-operation can also contribute to the advancement of science in areas where the need is greatest.

In fulfilment of the task assigned to it in respect of scientific and technological co-operation, Unesco exerts particular efforts to promote research and teaching in the various branches of the exact and natural sciences and in their applications.

The Organization's first Medium-Term Plan (1977–83) gave these efforts a strong impetus, facilitated by the growing interest shown by the international community in the possibilities which advances in science and modern technology were offering for development.

The second Medium-Term Plan (1984–9) adds a new dimension to Unesco's action in the present decade. Three of the fourteen major programmes which make up the Plan deal with science in today's world and in the world of tomorrow: Major Programme VI—'The sciences and their application to development'; Major Programme IX—'Science, technology, and society'; and Major Programme X—'The human environment and terrestrial and marine resources'.

It was therefore only right and proper that Unesco should make plans to issue a publication intended for the general public, taking stock of the experience it has built up in science and technology as applied to development. Such is the purpose of this book.

It is, in the first place, the culmination of a vast collective effort on the part of Unesco and the 400 or so non-governmental bodies which were associated with the Organization in the preparation of the current Medium-Term Plan. It is also a consequence of the work by an Advisory Panel on Science, Technology, and Society, a group of thirty-three specialists from all over the world, which met several times in Paris between 1981 and 1983.

Lastly, it represents a continuation of work commenced in two books previously published by Unesco which paint a general picture of scientific and technological progress and its consequences for humanity: *Current Trends in Scientific Research*, written by the eminent French physicist Pierre Auger and published in 1961, and *The Scientific Enterprise, Today and Tomorrow*, compiled by the late Italian geneticist, A. A. Buzzati-Traverso, published in 1977.

Ultimately, however, this book is the individual handiwork of its author, the well-known British science writer Robin Clarke, who is thoroughly conversant with the impact of scientific and technological progress on society and at the same time well versed in the operations of the United Nations system in this area.

Unesco enabled him to write the book but gave him complete freedom to express himself as he wished. Thus, all the credit for the work's major qualities—mastery of the subject, coherence of presentation, and clarity of style—must unquestionably go to him. By the same token, he bears sole responsibility for the case he presents; and while Unesco agrees with much of it, the Organization of course reserves the right to dissent from some of the views expressed.

This book should prove very useful to the many readers throughout the world who have been waiting for a comprehensive, accurate, and at the same time easy-to-read work on this subject. The fact that it is being published by Unesco in conjunction with a major publishing house outside the United Nations system, Oxford University Press, gives it every chance of meeting with the wide public acclaim it is entitled to expect.

Amadou-Mahtar M'Bow

Contents

In fond memory of
ADRIANO BUZZATI-TRAVERSO

Introduction

It were good therefore that men in their innocence would follow the example of time itself; which indeed innovateth greatly, but quietly, and by degrees scarce to be perceived.

Francis Bacon, 1561–1626

The reasons for the publication of this book have much to do with the fact that Bacon's wise advice, given some 400 years ago, has not been followed. In the industrialized countries, most of today's pressing problems are connected with science, technology, and innovation. As Michel Batisse, Unesco's Deputy Assistant Director-General for Science, has written:

Scientific research is under fire in many quarters of government and public opinion. There is a marked disenchantment, particularly in western industrialized countries, with the social and human benefits of science which, in common language, is generally but erroneously equated with technology. The threat of nuclear weapons, the pollution of the environment, the uncertainty of the supply of resources—all combined with inflation, unemployment and the tensions arising from rapid modifications of value systems—are seen as instances of failure of the conscious or unconscious dream that science and its proper application would solve the problems of mankind.[1]

Elsewhere the problems are quite different. In the developing countries, policy-makers deplore the lack of scientific development in their own countries and the lack of impact which glamorous advances made in the rich countries have on their own pressing problems.

There are thus two issues to be addressed, but they are interconnected. The aim of this book is to shed light on them, a task which can be done only by tracing the roles which science and technology have played in social evolution in the past, and hazarding some guesses as to how those roles may change in the future.

Inevitably, the analysis must be partial and selective. The filters which we shall use to sort the relevant from the irrelevant will be those of the United Nations Educational, Scientific, and Cultural Organization, by whose initiative this book is published. For more than three decades, Unesco has had a special role to play in the problems of science and society. While many, if not all, the members of the United Nations family are involved with the effects of technology on human development—on food production, health, and family planning, for example—only Unesco is charged with responsibilities for science *per se*. And, as its name implies, it is in a unique position to catalyse investigations into the interplay between science, technology, culture, and education.

However, this book does not reflect Unesco policy, nor the policy of any of Unesco's many committees and programmes. In fact, it is not intended to advocate any kind of policy. It seeks, on the contrary, to explore the range of views and options of both educated people and specialists on the most important issues affecting the relationship between science, technology, and society. Some of the views examined in the pages that follow— some of them at length—would be endorsed by neither Unesco nor the author. This does not make them any the less interesting.

However, the inclusion of some of these views has led a few pre-publication critics to argue that the book be classified as 'anti-science'. How readers wish to classify any book is, of course, largely a personal matter. But there is certainly no intention to be anti-scientific. There is a pressing need for an informed social critique of scientific activity. However, a critical and thoughtful examination of science—or, more accurately, its role in world development—must not be confused with an anti-scientific attitude. Are those who claim that cars consume too much petrol 'anti-motoring' or those who accuse the United Nations of bureaucracy 'anti-United Nations'?

The idea for this book arose during a period in which Unesco was preparing its Medium-Term Plan[2]—an outline of the programmes it would undertake during 1984–9. During this process an Advisory Panel on Science, Technology, and Society was formed, comprising thirty-three scientists from almost as many countries. Much of what they said and wrote is reflected in these pages, but they are not, of course, responsible for the book's content, which extends its net wider than did the brief for the advisory panel.

Nor is this book an isolated publishing phenomenon. In 1961 Unesco published *Current Trends in Scientific Research*,[3] written by Pierre Auger, Unesco's first Assistant Director-General (ADG) for Science. This was followed in 1977 by *The Scientific Enterprise, Today and Tomorrow*,[4] by the late Adriano Buzzati-Traverso, Unesco's ADG for Science in the early 1970s. Since these two books were written, ideas about the relationship between science, technology, and society have changed radically, a fact which is reflected in the title of this volume: *Science and Technology in World Development*.[5]

The books already mentioned provide some obvious insights into the development of our ideas about science, technology, and society. Auger's masterful summary of what was happening in science during the 1950s did not deal with the social implications of laboratory work. In those days, more than thirty years ago, few people, and even fewer scientists, were concerned about such things. These were the 'boom years' for science and technology. The economy was buoyant, and money for research was as easily available as the research was productive. Technological optimism was rampant. Phrases such as 'the final conquest of human disease'—which now seem absurdly optimistic—were commonplace.

By the time Buzzati-Traverso took a second look at the scientific enterprise, conditions had changed almost beyond recognition. These were the 'doom years'. The American biologist Paul Ehrlich was prophesying imminent disaster from countless directions. The environmentalists were pleading for a halt to development, and even to science itself. And Buzzati-Traverso faithfully reflected the mood of those times, which, for all its alarmist and revolutionary zeal, was also one of intellectual ferment. During the late 1960s and early 1970s the extremists forced many of us to take stock; to think again about the meaning of development, the nature of science, and the carrying capacity of our planet. Despite the false prophesy of the doomsters, their warnings were salutary.

We have now entered a third phase of the relationship between science, technology, and society. It is characterized most importantly by the urgent demands of the developing countries for a larger share in the results of scientific activity, and by those countries' determined efforts to build up their national scientific and technological capabilities. At the same time, the developed countries are displaying a more cautious attitude towards the roles of science and technology.

During the 1950s and 1960s the high rates of growth of spending on research and development appeared to be closely linked to high rates of growth in productivity in both the United States and Europe. A number of factors now call this relationship into doubt. First, recent studies suggest that even in the boom period those countries which spent most on research and development in relation to their gross national products—namely the United States and the United Kingdom—had among the lowest rates of growth of productivity. Second, the economic slow-down which has occurred since 1973 was not presaged by parallel decreases in research activity, except in the United States and France. As two US economists have put it, 'The experience of the 1970s casts doubt on the presumed tight link between a nation's overall R and D spending and its rate of productivity growth.'[6]

Today, 90 per cent of research is still carried out in a few rich countries. No one doubts that if that 90 per cent took place in the Third World, and were focused directly on the problems of development, the results would be substantial: both the pace and the nature of development would certainly change for the better. But catalysing endogenous science and technology in developing countries is not easy. It is often expensive and it is invariably slow. Even if successful, the means of linking the results with the economic and social needs of a country are frequently elusive—as indeed they have often been in the developed countries.

It is worth recalling, at this point, that the technology of today is already fully capable of providing a decent life for everybody.

We have not, so far, exceeded the carrying capacity of our planet; on the contrary, according to one of a number of reports issued by the United Nations Environment Programme to celebrate its tenth anniversary:

There is still a great potential to expand the ability of the global ecosystem to support more people, by judicious combination of making more productive and rational use of the Earth's available resources and by a more equitable distribution of the benefits of various production processes.[7]

Taken as a whole, then, our planet is rich enough to look after us all—to provide sufficient food, energy, housing, and employment. A change in priorities, and extensive political and economic reform, might solve development problems within a decade. To expect that to happen, however, is just as Utopian as it is to expect the magic wand of science and technology to solve the problem. Neither is sufficient, but both are necessary. And they will become increasingly so as populations increase, resources become scarcer and more expensive, and the demands of the underprivileged become more strident.

The second point which needs emphasis concerns the relations between science and technology. It is often argued that technology is produced only as a result of scientific research. This is not the case. All the societies which have ever existed on this planet have possessed a technology; very few of them indeed have possessed a science. Clearly, traditional or indigenous technologies are not, and never were, science-based. A decade or two ago they were in danger of rejection for that very reason. Today we have at last learnt that an indigenous technology may well prove a fertile starting-point, at least in rural development. But involving scientists and engineers in the process of adapting and improving traditional technologies is not proving easy.

Even in the industrial world, technology and science-based technology are two very different things. In fact, most of the things we use today were developed by a series of gradual improvements in design. Of those innovations which suddenly burst upon us, a great many were never born inside a laboratory. From the piano and the steam-engine to the zip-fastener and the ball-point pen, we have countless examples of technical innovation which stem more from the ingenuity of the human species than they do from scientific research. The steam-engine was not developed, as is often assumed, from an improved theoretical understanding of the principles of thermodynamics; this came only later (though the theory might have been long delayed were it not for the invention of the steam-engine). Innovation is one thing, the production of science-based technology another.

Yet it is true that during this century, and particularly since the Second World War, science-based technology has produced innovations which have had a social impact orders of magnitude greater than other forms of innovation. The introduction, for example, of nuclear weapons, nuclear power, electronics, and new data processing techniques has had, and is still having, social repercussions which go far beyond anything we have previously experienced. Other forms of innovation rarely produce such social upheaval.[8]

It would appear that only science-based technology has this built-in capacity to shock. Indeed, other forms of innovation, produced as they are from within the social context, and usually developed from existing artefacts, seem almost bound to produce gentler, more easily assimilated changes. This is why Unesco's programme of investigation is called 'Science, Technology, and Society' and not 'Innovation and Society'.

It is an important point, and one which is often overlooked. Today it has several implications. In the developing countries, there has been a tendency to equate progress with science-based technology. At best, this is a partial definition. Other forms of innovation are fully capable of improving the human condition, usually without the shock waves of change which can accompany the introduction of scientific technology. They should not be forgotten.

Happily, there is some evidence that they are being revived. The UN Food and Agriculture Organization, for example, is encouraging wood-rich developing countries to use fuel-wood as a fuel with which to promote industrial development. In that way, FAO argues, countries can make a 'soft transition' from traditional to advanced technologies without the social disruptions which often result from the widespread introduction of fossil fuels.[9]

Similarly, the business enterprise, looking for new products, too often searches only within the confines of scientific technology. Yet an analysis of the most successful businesses operating in the world would reveal many of them marketing products which outsell their competitors simply because they are superior and/or cheaper.[10]

So far, this discussion has focused on technology—the visible part of the science/technology spectrum. The urgency of the world situation is such that action programmes are the only ones currently likely to succeed in raising substantial funding, either nationally or internationally. But of course the equation of science with science-based technology is both a recent phenomenon, and an inexact one. Science itself was conceived and born for quite other reasons. By the eighteenth century it alone was claiming unique access to an objective understanding of the world, a claim which is continued to this day, despite fierce controversy led by several imminent philosophers of science. The issue of whether science is the only technique for gaining an objective view of the world is a fascinating one but is something which can only be touched on in this book. Once again, the lesson we have learned in the past decade is that such matters are complex and, even if science is no longer to be trusted as the sole arbiter of truth, it remains an immensely powerful—and probably the most powerful—means of imaging the world.

In truth, of course, science is a means of constructing models of reality. The relationship of those models to reality itself is perhaps unknowable, even though many of them exhibit remarkable predictive powers. At first sight, such a view of science might seem to detract from its appeal. And, indeed, when such criticisms first became widely debated in the 1960s they produced a stern response from the scientific community. In fact, time has shown that the citadel of science is not undermined by philosophical probing into its ultimate foundations. As scientists themselves know full well, an improved understanding of how anything works cannot but strengthen the nature of the appeals it may make.

One of the most contentious criticisms used to be that the

scientific viewpoint was blinkered by reductionism: a scientific insistence that reality could be approached only by viewing its constituents in ever smaller chunks—for instance, one should investigate the nature of matter through examination of the subatomic particles of which it is comprised, or try to understand the nature of life through the study of the chemical molecules of which it is composed. Encouragingly, while research of this kind is still probably the mainstream of scientific activity, the past decade has seen a notable resurgence of holistic thinking. Mainly as a result of the environmental movement, we have come to appreciate the remarkable links which join the world's natural systems together as a continuous whole. Studies of these interrelationships are an increasingly important part of scientific activity.

Nevertheless, science for science's sake—or, better, for the sake of improving our knowledge of the world—appears to be becoming less fashionable. The developing countries often claim that their time-scales for reform are too short to nurture such luxuries—we shall examine numerous instances of this approach, and its implications, in this book. And shortage of cash is forcing cut-backs in the developed world, of which basic science is taking a disproportionately large share. One result is renewed interest in an issue of great philosophical interest: are there limits to what we can know, or can we continue making scientific discoveries at the same pace indefinitely? If there are limits, it is appropriate to ask whether discoveries now arrive more slowly than they used to and whether they are now more expensive to obtain.

A contrary view holds that such questions are irrelevant. Science, according to this view, is a natural human activity, a part of human culture in the contemporary world. In this sense, science is to be regarded as a social activity and its funding should be no more controversial than subsidizing a public transport system or a national health service.[11] Another related view is that pure science/applied research/development is a continuum, and that to ignore one aspect of it is to imperil the others. Buzzati-Traverso put it succinctly: 'It must be conceded', he wrote, 'that in modern societies science and technology represent a continuum. No borderline between the two can any longer be drawn and, therefore, science is technology.'[12]

If this is true, the current trend towards applied research will surely turn out to be counter-productive. At one level, it may prove impossible to extract the fruits of scientific technology from a tree which is inadequately fertilized with a basis of pure science. At another level, the very notion of development used in this context may be too limiting: though basic needs are often regarded as defining the thrust of development efforts, those needs are not purely material. As Johan Galtung has argued,[13] the needs of the mind and the body are never so neatly divided as our development programmes seem to insist. In short, is science a cultural luxury or a social necessity? Time will tell, but answers to such questions are urgently needed.

These, then, are some of the issues to be discussed in Part I: *Concepts*. From such generalities it is important to turn quickly to the real world. We are not, in fact, concerned with the problem in theory but in practice: what is likely to happen within the next ten or twelve years which will affect society? Part II deals with *Trends*.

The first areas to be considered are the breeding grounds of change—the basic sciences. It is here that tomorrow's technologies may be conceived, and we look briefly at the state of the art in biology, chemistry, and physics. Several things are immediately obvious. While physics has held the stage for the past fifty years, its role as the cutting edge of scientific advance is now being steadily eroded by biology. The life sciences are providing us with new knowledge about the nature of life which is already being translated into practical action. They are also showing us how to eke out precious reserves of energy in the form of biomass, and convert them to usable products. Of course, plenty of surprises may be in store from physics and chemistry; but when the life sciences have so much to offer the contemporary world, plus of course an improved understanding of the interrelationships between the different sectors of the environment, it is not difficult to predict biology's continuing popularity. Especially, as states-men are quick to observe, when it is also so much cheaper than is research at the frontiers of the physical world.

A decade ago, one might have proceeded direct from here to the technologies themselves. But no longer. Since the historic United Nations Conference on the Human Environment, held in Stockholm in 1972, our view of the world has changed radically. Whatever we do on this planet alters it, either for the better or for the worse. Thus far these alterations have been made largely unwittingly. It is time we understood what we are doing, and world-wide efforts, orchestrated by the United Nations Environment Programme, are under way to help us do so. This has created a new need: a need for something which is neither pure science—aimed at increasing our understanding of the natural world—nor technology, aimed at controlling it. We now need to understand what we are doing to the world, and how our actions in one area affect other, apparently quite remote areas.

For these reasons it is important to review progress in what we have called the natural systems—the Earth's crust, the land, fresh water, the oceans, and the atmosphere. As we shall see, reports from some areas are encouraging: the oceans, for example, have not been grossly polluted except near coasts; and while their yield of edible fish is not expected to increase substantially, mineral production is. The news from other areas is less happy: dire things are happening in the atmosphere, to our fresh water supply, and to our productive land. As our understanding grows, our worries increase—though, it must be stressed, not in the cataclysmic sense of the early 1970s. None of the disasters so confidently predicted a decade ago have arrived, nor seem likely to. But there is much to be concerned about.

In turning to the technologies themselves, we have of course a double-edged sword to consider. In a very real sense, technology marches on regardless of preferred directions. Satisfactory control systems do not yet exist. Technology—particularly when financed for power or greed—has a momentum of its own, and national and international legislation to control it remains relatively impotent. Lord Zuckerman, describing how science led to an understanding that splitting the atom could release vast amounts of energy, recently had this to say of the resulting arms race: 'From that moment technology assumed command. A new future with its anxieties was shaped by technologists, not because they were concerned with any visionary picture of how the world should evolve, but because they were merely doing what they saw to be their job.'[14]

Those anxieties, of course, have not gone away.[15] The debate about nuclear weapons, which has now been in progress for five decades, intensified sharply during the early 1980s. Criticism of spending on arms in general was, if anything, more vociferous than it had ever been. And that criticism was no longer confined to the industrial nations. The developing countries are doing themselves and their populations untold harm by pouring vast quantities of their precious resources into armaments, a form of investment which they should have learned from the developed countries is financially a bottomless pit.

Important though the military preoccupations are, they are not the only areas of technological concern. Improved data processing is putting people out of work; the life sciences will undoubtedly provide techniques which we may find more of a problem than a solution. Will we, for example, benefit from being able to choose the sex of unborn children? At the same time we are in dire need of improved technologies for food production, energy production, and health. Those technologies are round the corner. Even now we can discern something of their future shape.

But as technology is not a virtue *per se* (as some argue that science is), the future of technology involves much more than simple prediction of what we may or may not be able to do. We must also consider needs. What technologies are actually wanted —and how urgently? Recently, several attempts have been made to draw up a programme for technical change.[16] To compare that programme with what seems likely to arise from planned research is instructive: not only does it reveal the gaps—which we may well have to make efforts to fill—but it highlights those areas which are carried on by their own momentum. How, then, do we alter the focus of research, steering it nearer what we need, and deflecting it from ends which seem more dubious or less beneficial? Even if we knew how to do this, who should define what we need and what we don't need is becoming one of the key political questions of the day.

This brings us to Part III: *Choices*. The *laissez-faire* model of science and technology has now been dead for several decades. But the decision not to let a horse run wild is a different thing from being able to direct it where to go. In this case our problems are compounded: not only must we wrestle with the general problem of control, but we have to resolve where it is we want to end up. Each individual's private view of Utopia could be matched to his own research programme. That way we might end up with 4,500 million different sets of research priorities.

Such problems can be easily dismissed by stating that this is a field for due political process—national science policies will take care of the problem. There is now abundant evidence that this is not so. For one thing, for the first time ever, the general public has begun to express its own opinion forcibly[17]—so forcibly, in fact, that nuclear power programmes in several countries have been reduced to a state of stagnation. In many ways, this has been the single most significant development in the interaction of science, technology, and society in the past decade. It seems safe to predict that public concern will make itself felt in other areas in the coming years. Neither governments nor their science advisers can any longer have it all their own way. This is a lesson which will take some digesting.

However, areas of public concern are not arrived at simply. The media have a role to play in formulating opinion. So do scientists, as they are becoming increasingly aware. In turn, these actors interact with others, including national policy-making institutions, the large businesses, and particularly the multi-nationals. So we must look in turn at the roles played by the scientific community, the nation state, and the multinational enterprise in formulating science policy, and in effecting action.

Were this the end of the story, there would have been little need for Unesco to be devising science and society programmes for the next decade. But, of course, it isn't. 'Only one earth' became the motto of the Stockholm conference, and its message applies as much to science as it does to the environment. The international component of research has always been important: as science becomes more expensive, and as increased efforts are made to encourage endogenous scientific development in the poorer countries, it will become even more so.

In this area, Unesco has had a number of major successes. It was early on responsible for the setting up of CERN—the European Centre for Nuclear Research—which quickly became one of the world's leading laboratories in the investigation of the nature of matter. More recently, the International Centre for Theoretical Physics in Trieste has provided a place for scientists from all countries to carry out their research. And on the environmental side, Unesco's Man and the Biosphere Programme has for many years been directing attention to the human environment.

Strangely, as the urgency for poor countries to begin their own endogenous scientific development increases, the international component of science becomes more important, not less. No one believes any longer that a developing country can simply import its science and technology (for reasons which will be dealt with at length in this book). But equally the development of a scientific base within a country cannot be initiated or stimulated simply at national level. Assistance is always needed—in planning, in finance, in training and education, and in general policy. Such assistance can be provided bilaterally but an international approach is often more useful.

Nor can a developing country develop its scientific resources unless those resources are themselves part of the international scientific effort. Science cannot exist in isolation—each part of the system feeds off every other, resulting in cumulative advance. Isolationism in science is not simply counter-productive; it is impossible.

During the next decade, even more important issues may raise themselves. For two decades international collaboration in science has proved effective; in some cases, where research has been particularly expensive, as in particle physics, international laboratories have become the norm rather than the exception. Now that the focus of attention is changing, we must consider whether international laboratories could be equally successful in tackling development-related issues such as nutrition and health, sanitation, and rural energy. To be sure, the original motivation —that of the high cost of the research—is missing. Yet international or at least regional attacks on specific key problems might well save a great deal of repetitious and even poor-quality research from being done simultaneously in twenty or more developing countries. Needless to say, institutes of this kind would have to be located *in situ* in developing regions, rather than in Europe as they have been in the past.

Finally, of course, the issues discussed in this book will continue to be the subject of study for many years to come. These issues are not national ones; they are international. As the scientific resources of the developing countries become stronger, they will be drawn increasingly into the international debate on the role of science and technology in society. If Unesco can provide a forum for that debate, this will be as important an achievement as any it could make.

Part I · Concepts

1 Science for knowledge

One thing I have learned in a long life: that all our science, measured against reality, is primitive and childlike— and yet it is the most precious thing we have.

Albert Einstein, 1879–1955

Science, according to the *Concise Oxford Dictionary*, is the pursuit of systematic and ordered knowledge. As far as it goes, this will do as well as any for a definition. Yet there are almost as many definitions of science as there are scientists; and there are certainly as many as there are philosophers of science.

For some, the essence of science is objectivity; for others it may be creativity, reliability, its public availability, the fact that it is a social activity, or a revolutionary one, or, indeed, a cultural one. Science is all these things; and perhaps it might be thought that in a world where human well-being is so obviously and closely related to scientific and technical knowledge, it is not important to enquire into exactly why scientific knowledge is so valuable or into how it differs from other forms of knowledge.

Yet this kind of scientific introspection is becoming increasingly common, and not only in Western society. The reasons are not hard to find; in a world wrestling with both gross inequalities and a dizzying pace of social and technical change, there is a need to take one step back and ask where it is we are going. In so doing some people are questioning assumptions which have gone unquestioned for centuries. These assumptions all relate, in one way or another, to the role which science and technology now play in society, and to the role which they will play in the future.

In this chapter we are concerned with the most difficult of all the science/society issues—and the one which raises the fiercest emotions. That issue concerns the value of scientific knowledge. It was once fashionable to side-step that issue and simply to claim, as the positivists did, that science is the naked truth. For better or for worse, neither scientists nor the philosophers of science any longer believe this to be a profitable position to take.

Instead, they seek to ask increasingly pertinent and relevant questions. As the British physicist and philosopher John Ziman has put it, we must stop thinking in terms of simple yes or no answers to questions like 'Is science to be believed?' Instead, 'We should ask to what extent we should believe in science; we should enquire into its practical limitations; we should inform ourselves carefully as to the credentials of competing systems of belief; and we should explore cases where a scientific consensus is still in the making, or where unsettling doubts have been expressed.'[1]

If the details are complicated, the basic issue is at least fairly clear: in the West knowledge has traditionally been considered an absolute virtue and freedom of enquiry has been regarded as a scientific right. Since the Second World War that right has been eroded first by the State, which has sought to direct the attention of science to increasingly concrete areas, and more recently by concerns about how that knowledge may be used or what social implications its existence may have. Society may now be on the brink of accepting the idea that there are limits to scientific enquiry—and hence that there are social values of greater importance than freedom to generate new knowledge.

The decline of confidence

Let us start, however, with some facts. The annual average growth rates of financial support for academic science in the United States since 1960 were as follows:

1953–60	12 per cent
1960–68	14 per cent
1968–74	zero
1974–78	4 per cent

There are plenty of other yardsticks which indicate the same sort of trend: for instance, the number of Ph.Ds. produced in the sciences and in engineering in the United States fell from some 14,311 a year in 1971 to only 11,777 by 1977; the percentage of grant applications funded by the US Institutes of Health fell from 82 per cent in 1967 to only 48 per cent by 1977; and similar trends could be observed in most European countries.

The net result was dramatic. As John V. Granger writes:

The adjustment process took a traumatic toll. Whole laboratories were shut down or pruned beyond recognition. Research scientists and engineers were forced to find new employment outside the R and D enterprise, many of them never to return, and university enrolments in scientific and technical curricula declined drastically. Science ministers were dropped from cabinet or subcabinet status or disappeared altogether from the political scene in many countries.[2]

Why? Part of the reason, of course, was economic. Even so, in times of hardship we cut down on what we value least and for the previous decade or two it would have been quite fair to claim that science was one of the things we valued most. So why the vicious reaction to science? Almost by definition, the facts were that those who gave money for science, and the general public whom they represented, had become disenchanted. A survey made by the National Science Foundation in 1974 showed that 39 per cent of the American people who expressed any opinion did not agree with the proposition that 'Overall, science and technology do more good than harm'. As Ziman pointed out:

The attack on science comes from many quarters, but is not well concerted. The medley of opposition includes many strange companions-in-arms, following contradictory causes. The conservative fears that science will destroy the only world that he knows; the progressive imagines that it will poison the paradise to come; the democrat is cautious of the tyrannous capabilities of technique; the aristocrat fears the levelling tendency of the machine. The pleas of defence are equally inconsistent: some say that scientific progress is automatic and inevitable; others that the future must be determined by rational scientific planning; technocrats delight in telling us that science will make life more comfortable; space addicts proclaim that man must go forth and conquer the universe.[3]

The scientific establishment took recourse in one characteristic line of defence: it began to claim that attacks on science, and the cutbacks which followed them, were 'irrational'. Thus in the

United States the President of the National Academy of Sciences was quoted as saying: 'Particularly troublesome is the ever more frequent expression of the notion that there are questions that should not be asked, that there are fields of research that should be eschewed because mankind cannot live with the answer. Nonsense. No such decision can be rational, much less acceptable.'[4]

This type of reaction has, unfortunately, served mainly to conceal the matters of real import which underlie the contemporary situation. Though it is true that some reactions against science are irrational, we shall be dealing here with what must be classified as rational objections. As Loren R. Graham, Professor of the History of Science at the Massachusetts Institute of Technology, puts it: 'Many of the concerns about science and technology are legitimate, in the sense that they represent genuine fears for the safety, morality, and the sense of human purpose and worth of society. If we dismiss all anxieties about science as "irrational", we will not be listening to some important debates.'[5]

It helps, of course, to analyse these anxieties carefully for they spring from a variety of disparate roots. Most of the objections ever made can be fitted somewhere into the following format, which is an expanded version of one first published by Graham:

i. Concerns about science

- science diminishes other forms of knowledge (*single-vision science*)
- knowledge itself can be subversive (*subversive knowledge*)
- knowledge leads inevitably to technology (*inevitable technology*)
- science can be used to incite prejudice (*prejudicial science*)
- research on humans or animals is unethical (*objectionable research*)
- accidents may happen during research (*dangerous research*)
- the cost of research is high (*expensive research*).

To these must be added the better-known and more frequently examined concerns about technology:

ii. Concerns about technology

Technology may:

- damage the environment or man (*destructive technology*)
- produce difficult ethical choices (*slippery-slope technology*)
- be used for economic exploitation (*exploitive technology*)
- alienate the individual (*alienating technology*)
- waste resources (*wasteful technology*)
- need large capital resources (*capital-intensive technology*)
- destroy local cultures (*culturally destructive technology*).

Comparing these two lists, it is immediately obvious that the concerns about technology are more easily countered than are the concerns about science. With one exception (*slippery-slope technology* above), the concerns about technology all depend on the type of technology used; and hence they can be countered by changing the nature of technology. This, in essence, was what the appropriate-technology debate of the 1970s was all about.

However, there is as yet no such thing as appropriate science (though some suggestions have been put forward[6]). The concerns about science must therefore be met head on, and it is worth describing each of them briefly (as well as slippery-slope technology, which also poses some fundamental issues).

Single-vision science. During this century the words knowledge and science have become almost synonymous in the industrialized countries. One result has been a diminution in the value attached to other means of acquiring knowledge. In fact, of course, knowledge can be acquired in many different ways, of which scientific research is only one. (So can the closely related concept of wisdom; the Three Wise Men pre-dated the Enlightenment by nearly two millenniums.)

However, the fact that only science claims to provide an objective approach to the truth gives it a special position. This position has been somewhat eroded of late, as philosophers of science have come to understand more clearly the reality of the processes by which science advances.[7] The relationship between knowledge and science, and their relationship to such things as wisdom and spiritual experience, is still a largely unexplored field and one which may well repay study during the 1980s. It is both important that scientific knowledge be treated with the respect it deserves, and that it should not be elevated to a monopolistic status with regard to other branches of knowledge and learning.

The idea that scientific knowledge has a limited (though immense) relevance, and that other forms of knowledge are equally admissible (though less useful), has a long history—such ideas were fiercely defended by Goethe and Blake, and more recently by Theodore Roszak in *Where the Wasteland Ends*.[8] However, many distinguished scientists have maintained similar ideas. In the 1920s Sir Arthur Eddington spoke of 'using the eye of the soul' to obtain knowledge, and of forms of knowledge which were not accessible to scientific method.[9]

This idea that there are other forms of knowledge which are as admissible and valuable as science has, for some reason, often been regarded by scientists as an attack on their craft. This is not the case. On the contrary, the idea itself does not diminish science but in some ways enhances it because it focuses attention on the special qualities of scientific knowledge which set it apart from other forms of knowledge. On this point it is worth quoting the American philosopher Ruth Nanda Ashen at some length:

in viewing the role of science, one arrives at a much more modest judgement of its function in our whole body of knowledge. Original knowledge was probably not acquired by us in the active sense; most of it must have been given to us in the same mysterious way we received our consciousness. As to content and usefulness, scientific knowledge is an infinitesimal fraction of natural knowledge. Nevertheless, it is knowledge whose structure is endowed with beauty because its abstractions satisfy our urge for specific knowledge much more fully than does natural knowledge, and we are justly proud of scientific knowledge because we can call it our own creation. It teaches us clear thinking, and the extent to which clear thinking helps us to order our sensations is a marvel which fills the mind with ever new and increasing admiration and awe. Science now begins to include the realm of human values, lest even the memory of what it means to be human be forgotten.[10]

Subversive knowledge is the area of traditional objection to science. Galileo's ideas were attacked by the Church of the time on the grounds that they were demeaning to Man because he was no longer placed at the centre of the universe. Darwin's theories were attacked for similar reasons. In the past, established ideas have often been held more important than the results of new research, which have then been suppressed for religious or political reasons. We generally assume that the issue of subversive knowledge is of historical interest only.

In fact, this is not the case: a number of contemporary issues fit into this category of ideas. For example, many studies now purport to show that human behaviour can be related to that of animals, and explained by reference to them. Humanists who might otherwise favour scientific explanations often tend to reject such ideas simply because they regard them as degrading to man; they argue that it is the differences between man and the other animals, and not the similarities, which are interesting.

Similarly, a number of scientists have questioned whether it is wise to attempt to communicate with extraterrestrial civilizations because if they exist, and if they are more advanced than our own societies, the effect of contact with them might well be as catastrophic as has been the impact of civilized man on the primitive peoples of this planet. As Robert Sinsheimer has said: 'In my view the human race has to make it on its own, for our own self-respect.'[11]

Inevitable technology will be discussed in the next chapter where the relationship between science and technology is examined in more detail.

Prejudicial science merits serious consideration here. For one thing, the fact that research can be used to provide scientific evidence in favour of prejudice—whether it be for one race, one class, or one sex—helps explain the anomalies observed under the heading of subversive knowledge. Indeed, the distinction is far from clear; many scientists and non-scientists take active steps to discredit any research—such as that on intelligence and race—which could be used to reinforce prejudice (thus, some would say, exhibiting their own prejudices). They do not do so for scientific reasons but for moral ones—and their actions, even if not very scientific, are often regarded sympathetically. However, such actions do in fact have much in common with the Church's rejection of Galileo and Darwin, which is today generally condemned. Perhaps surprisingly, the values we assign to science, research, and knowledge frequently turn out not to be self-consistent when they are examined critically.

One of the major problems connected with prejudicial science is the stamp of authority which scientific pronouncements carry. This authority differentiates scientific activity from scholarship in other fields where the conclusions reached are often expressions of personal opinion, albeit ones which are formed after extensive study. Thus when a scientist announces research results which may have social implications, the announcement is likely to be widely accepted. If these ideas later turn out to be wrong, as often happens in cases of prejudicial science, the harm has already been done.

Many scientists are aware of this danger. Dr Philip Siekevitz of Rockefeller University pointed out the dangers well over a decade ago:

I think our greatest sin is to presume to know much more than we do, and even if we don't, we give the impression that we do, and so the world takes our tentative findings and makes them actualities.[12]

Objectionable, dangerous, and expensive research is intellectually less interesting simply because science could continue, and remain immensely valuable, without being objectionable, dangerous, or even very expensive. However, all these issues raise the question of how highly we value knowledge, and they illustrate the fact that in practice we rarely regard the desire to obtain knowledge as an absolute right; in practice we curtail it heavily. Few people indeed will admit that it is right for knowledge to be obtained at the expense of human suffering caused by experiments on human beings. Fewer and fewer people believe it is right to subject animals to painful experimentation for the sake of knowledge. And hardly anyone believes that an individual researcher has the right to conduct his or her own research if there is a risk of accident to others.

Slippery-slope technology, the final issue to be looked at here, illustrates the ethical complexity which can be brought about by technical change. For example, if technology develops life-support machines, are people who 'pull out the plug' murderers? And if we don't define them as murderers are we not half way down the slippery slope of condoning the actions of someone who suffocates a deformed child? By altering the range of ethical choices we may have to make, new technology also threatens to change the nature of those ethical choices—and hence to alter our ethical base.

Although medicine has always had to deal with difficult moral dilemmas, the possibilities of new biomedical techniques have compounded them many times over. In fact, so specialized has this field become that a number of institutes, such as the Kennedy Institute Center for Bioethics in Washington, DC, have been created specifically to study this field. One positive result of all this is that the study of ethics, in decline for more than a century, has had a rebirth. The subsequent resumption of moral dialogue demands this time that the scientist involve himself in the ethical issues he has helped to raise.

Science and values

These examples of current concerns about science and technology all relate, in one way or another, to the question of values. How science, and scientists, deal with the ethical, political, and ideological issues involved is thus of crucial importance. According to Loren Graham, there are basically two positions to take with regard to science and values. One school of thought—which Graham calls restrictionist—regards the world of values and the world of science as two distinct spheres, with no overlap. Restrictionists do not believe that the results of scientific research can have any bearing on value systems: they believe in science-free values.

Restrictionists would therefore deny that science has anything to say about the Biblical account of creation, about the nature of God, or indeed the nature of man. As we saw, Eddington was a restrictionist, believing that the world of the spirit and the world of science were quite separate. So was Einstein, who berated those of his colleagues who tried to show that modern physics was relevant to ethics. He believed that facts and values could not be compared; 'Science can only ascertain what *is*,' he said, 'not what should be.'[13]

Society, of course, can feel safe from restrictionists, for their science does not threaten to change social values; indeed, it has absolutely nothing to say about values. The bargain, however, is mutually beneficial. Science is equally free from censure by society for the very reason that it is held to have no political or ideological relevance, although it will of course have major material implications. In a restrictionist world, social ideas cannot be subjected to scientific criticism and scientific ideas cannot be subjected to social criticism. In a sense restrictionism provides mutual protection for both science and society.

The second school of thought, known as expansionism, is the opposite. Expansionists believe that science does indeed have things to say about social, political, and religious values. The following statements are all examples of expansionist thinking:

- studies of the structure of matter indicate there must be a Supreme Architect (God) [This view was common in the nineteenth century]
- Darwin's theories support the need for capitalism (social Darwinism)
- animal studies indicate limits of human behaviour
- behaviourists argue that values are nothing more than the effects of reinforcement.

Clearly such people as Jacques Monod, Konrad Lorenz, Desmond Morris, B. F. Skinner, and E. O. Wilson (the proponent of socio-biology) are all expansionists. So were Niels Bohr and Teilhard de Chardin. Adriano Buzzati-Traverso's statement 'Only by forcing the world of politics to accept the international ideals of science can we redeem our activities, and claim again that scientific endeavour is one of the noblest activities of man'[14] is a supremely expansionist statement.

Many attempts to justify the expansionist view rest ultimately on the philosophy known as reductionism—the idea that everything in the universe can be explained by reference to laws operating at the atomic and subatomic level. The physics of the 1930s appeared to give a tremendous boost to such ideas but even more important were the findings of the molecular biologists in the 1960s, who began to claim that they could explain all life in molecular terms. Foremost among the advocates of reductionism at that time was Francis Crick, who in his famous book *Of Molecules and Men* wrote that the crucial difference between a rock and a virus was the 'amount of ordered complexity, all at the atomic level'.[15] While this may be true to a reductionist, the differences between rocks and viruses are, to most people, far more obvious and a great deal less subtle.

In its purest form reductionism holds that ultimately everything can be explained in molecular or atomic terms—even the workings of the human brain and the nature of human emotions. Thus human values themselves can also be explained scientifically —when enough research has been done to understand the exact molecular functioning of the universe. This highly deterministic philosophy, of course, carries with it an attitude which often appears extraordinarily arrogant; when Crick lamented that many people still did not feel that science really involved them, he warned that science will soon 'knock their culture right out from under them'.[16]

Since the 1960s events have not gone the way of the reductionists. Holistic thinking came back into fashion during the environmental crisis of the 1970s and reductionism came under heavy fire from many scholars. However, it is perhaps important, as Graham warns, to distinguish carefully between reductionism as a research programme and reductionism as a philosophy. The former is clearly of great utility; and while science in the past decade has shown a remarkable and until recently little appreciated ability to deal with events and phenomena in a non-reductionist way, it remains true that reductionism has been the mainspring of scientific research for a very long time; it is, in fact, a necessary part of science though certainly not the only one.

The philosophy of reductionism is a different matter, as Graham warns:

reductionism as a philosophy is not merely incomplete; it is positively dangerous, for it dismisses the worth of older value systems as clearly unscientific while it provides no adequate substitute for them . . . unscientific though they are, based on religion or superstition though they may be, [value systems] act as temporary restraints that play critical roles.[17]

Part of the contemporary problem may be that although the extremes of reductionism are not now regarded sympathetically the solution of retreating into the comforting world of restriction-ism no longer seems to be a viable alternative. It has become so clear over the past decade that the advance of science and technology *is* informing our values that to return to a position where we continue to deny this may now be impossible.

The French scientist André Danzin believes that what is in progress is of great importance—a kind of reordering of the scientific mind in which the old and discarded concepts are chucked out and the new, more liberal components of scientific philosophy adopted. He writes:

Scientific thought is now particularly concerned with complexity. It is rejecting, at least temporarily, all simple and mechanistic explanations of the world. It is renouncing scientism, rediscovering humility, reducing rationalism and Cartesianism to the rank of auxiliary tools of reason and judging them to be insufficiently profound.[18]

Just as we learnt in the 1970s that science is not value-free, so we must accept in the 1980s that our values are not science-free. The implications of this idea for the future of the science/society debate are profound. It has led, for instance, to the need to establish just how highly society values knowledge.

Are there limits to scientific enquiry?

When the American biologist Sinsheimer asked whether it is proper for us to try to make contact with other civilizations which may be more advanced than our own, he went on to question whether research in two other areas will really be beneficial to man: one is the improved fractionation of isotopes, making it easier and cheaper to obtain nuclear weapons (an example of destructive technology); the other is research on ageing. The implications of the latter, Sinsheimer says, are clear: 'In a finite world the end of death means the end of birth. Who will be the last born? If we propose such research, we must take seriously the possibility of its success.'[19] Here is an example of research which could well complicate the meaning of existence almost beyond imagination.

It is important to stress that what is in question is not the control of technology—which has been a familiar issue now for more than a decade—but the control of research itself. The US National Research Council's view is that 'knowledge is better than ignorance and that it is better to regulate applications than knowledge'.[20] For many people, and for some scientists, this no longer goes far enough. As Sinsheimer asks: 'Where is the balance, the necessary check to the force of scientific progress? Is the accumulation of knowledge unique among human activities— an unmitigated good that needs no counterweight? Perhaps that was true when science was young and impotent, but hardly now. Yet we lack the institutional mechanisms for regulation.'[21]

Sinsheimer is a biologist and it is important to note that the current ideas about regulation of research do not all come from outside the scientific community. Indeed, in the case of experiments with recombinant DNA, it was the biologists themselves who first called attention to a possible need for restraint in research. An extreme view was given by the professor of chemistry at Princeton University, Kurt Mislow, in 1977. Addressing the problems of research with recombinant DNA, he said:

I will undoubtedly provoke cries of inquisition and the like, but I must nevertheless force myself to say that I don't agree that freedom of inquiry should be limited only if actual hazards are perceived. I do not agree that increased human knowledge is of paramount importance. I do not agree that the real enemy is ignorance. I think these are trademark shibboleths which everybody accepts without questioning. I can think of lots of examples where knowledge is extremely dangerous. And in the search for knowledge, you have to ask what you are going to do with the knowledge once you have acquired it.[22]

This is remarkable enough. What is almost equally remarkable, though, is the role which the public has played in this debate—a break with tradition which is so complete that it is the subject of a special chapter in Part III (see Chapter 9 'The public and the media'). For the first time in the history of science the public is seeking methods of direct intervention in the formulation of scientific research programmes.

Declining research budgets, of course, are also a direct reflection that a declining value is being attached to new knowledge. The key issue here is the value of knowledge. Contemporary society values knowledge highly—but, as in the past, not necessarily above all things. How we decide to value knowledge in the future is likely to have an important bearing on the health of the scientific community and on the kinds of regulation that are introduced to protect society from the effects of indiscriminate scientific and technical advance.

As yet, however, few people have stopped to ask what it is that informs our values about knowledge. Why do we value knowledge so highly? Do all societies? Does it seem likely that some cultures will value scientific knowledge less highly even though they value traditional knowledge a great deal? One of the few contemporary scholars to examine such questions seriously is Syed Muhammad al-Naquib al-Attas of the National University of Malaysia in Selangor. He believes that Western culture contains an internal tension which produces an insatiable desire to seek, born of a profound scepticism which 'knows no ethical and value boundaries'. In his paper *The Westernization of Knowledge*, he writes: 'The quest is insatiable and the journey perpetual because doubt ever prevails, so that what is sought is never really found, what is discovered never really satisfies its true purpose.'[23] According to this view, change, development, and progress are concepts which result from doubt, inner tension, and scepticism.

The issue of control

These philosophical matters are likely to be increasingly debated during the 1980s. But they are also accompanied by urgent practical issues. All the debates that have been, and still are, raging about science and values, and limits to scientific enquiry, can be seen as part of a much broader trend: the increasing control of science by society. In one sense this control has always been present—society has always determined, in one way or another, whether there should be more science or less science. But it has not, until the past decade, played much of a role in determining 'what science?'—a job which traditionally has always been left to the scientists, on the grounds that science has an internal logic, and new directions for research are revealed only in the laboratory.

Over the past fifteen years, that idea has been progressively eroded and more subtle concepts have come into play. The key words here are 'targeted knowledge' and 'directed research'— both terms which do not fit happily into the old linear concept of basic research, applied research, and development. This is a more significant development than is commonly realized: it amounts, in fact, to a change from the previous system (where there was a policy for science) to a completely new one (where we have science in support of public policy).

It has to be admitted, however, that the results have not so far been very encouraging. Certainly these new ideas have produced major discontent in the scientific community. To make matters worse, the change has not apparently been accompanied by major economic or social spin-offs.

However, as Granger writes, 'The reasons for the apparent failure to meet the new challenge so far are several. Science and technology policy makers have generally been preoccupied with the input (funding) requirements of the R and D enterprise rather than with the delivery systems that transform the research results into technologies and products serving real needs.'[24] It is to this subject—the relationship of science to technology—that we turn next.

2 Science for technology

The application of science to practical life has always been faced with the greatest difficulties and is even now, when its value is beginning to be recognized, carried on in the most haphazard and ineffective way.

J. D. Bernal, 1901–71

Scientific research has two products: knowledge and technology. The relationship between them was summed up nearly four centuries ago by Francis Bacon as 'knowledge is power'. But it is a great deal more complex than that. As everyone now knows, knowledge is not power—although, as David Dickson has pointed out, in the modern world control of the creation, acquisition, and use of knowledge can very often lead to power.[1] However, technologies do not arise only from research. In fact, for most of history, technology has not been derived from science or from research but from accident, trial and error, or indeed just plain hard thinking.

Most of the technologies familiar to man for the ten millenniums of his existence are of this kind—the discovery of fire, the invention of agriculture, the wheel, and the windmill, the domestication of animals and the production of most weapons, even including gunpowder, do not owe their origins to laboratory work. We sometimes forget in this scientific age that most of the societies which have ever existed on this planet have possessed a technology but not a science. And that most of the technologies most of us use every day are still more closely related to traditional crafts than they are to high technology—consider the table and chair, shoes and socks, and knife and fork.

The intimate relationship which science today enjoys with technology is a much more recent thing. Although it was arranged in the eighteenth century, the marriage of science and technology was not really consummated until well into the nineteenth century. Even then, the union functioned mainly in the field of chemistry and almost exclusively in the one country of Germany. Since then, the couple have grown together, though not always in the most productive ways. Science still has a habit of producing a technology which is not necessarily wanted; and, perhaps more serious, means have yet to be found of catalysing science into producing all the technologies that are badly needed. As we shall see, the latter problem is far from solution.

Contrary to popular belief, the industrial revolution was not based on science; and James Watt's steam-engine preceded rather than followed any improved understanding of the laws of thermodynamics—in fact, thermodynamics owed a great deal more to the steam-engine than did the steam-engine to thermodynamics.

It was not until after the Second World War that the general relationship between science and technology began to assume paramount importance; and the new science-based technologies which began to appear proved themselves capable of producing social change far faster than had ever been the case with traditional or indigenous technology. They also showed themselves capable of producing culture shock, and thus of producing effects on society which have no parallel in the history of innovations based on the slow and gradual improvement of traditional technologies. This should not surprise us; after all, a tradition is almost by definition something already assimilated by society and most of what emerges from the laboratory is not.

The linear model

The classic or linear model of the relationship between science and technology runs something like this. According to the Baconian tradition, scientists in their search for knowledge inevitably produce information which can be used to generate new technologies. It is then up to those working in applied research and development to produce the technologies and to the planners to see that the necessary capital, skills, and information are made available for the technologies to be put to use.

The role of science in all this is particularly interesting; according to the linear model, fundamental research has its own internal logic, which must always be followed, without regard to the possible technological spin-offs. Undirected research of this kind will sometimes produce a very handsome return in terms of the money originally invested, even though it is not possible to predict in which areas those returns may arise. For example, one famous study showed that research in agriculture paid dividends over periods as long as forty-five years of more than 700 per cent per annum in some areas, and even as much as 290 per cent per annum in a developing country such as Mexico.[2]

The kinds of argument used to justify such claims include the fact that science-based companies grow much more quickly than others. For instance, in the United States between 1945 and 1965 the economy grew by 2.5 per cent a year. Yet the average annual growth rates of the largest research-conscious companies were:

- Polaroid: 13.4 per cent
- 3M: 14.9 per cent
- IBM: 17.5 per cent
- Xerox: 22.5 per cent
- Texas Instruments: 28.9 per cent.

Similarly, as a result of this link between science and technology, it is often claimed that the length of time between discovery and application has been rapidly diminishing. For example, the gap between the discovery of polystyrene in 1840 and its application in 1936 was ninety-six years; yet by the 1950s the interval between the discovery and the application of polypropylene and the polycarbonates was only two to five years.

However, the linear model of the science/technology relationship has not stood the test of time. The three arguments advanced above are all fairly easy to refute. For example, it is not only science-conscious companies which have expanded rapidly. In some countries, shipbuilding and the property market have grown even faster than firms such as Xerox and Texas Instruments. And the studies purporting to show how the gap between discovery and application is now lessening appear to be based on selective sampling; the opposite case can be made with different examples, such as a two-year gap between the discovery and application of synthetic dyes in the 1860s and a five-year gap for the vulcanization of rubber at about the same time.[3] And, some commentators began to ask, if returns on undirected research were really that high, how much higher might they be if the research were actually aimed at achieving something concrete?

A chicken-and-egg problem

The linear model also began to be eroded from the other end. During its heyday it had been used to imply that the existence of an active scientific community in a country was almost sufficient to ensure rapid economic growth from the introduction of new technology. As the US National Academy of Sciences recently pointed out:

The belief that scientific preeminence goes hand in hand with techno-logical leadership is clearly false. While Japan's scientific stature was until recently not comparable to that of the United States, that country is nevertheless challenging us in many areas of high technology. Scientific excellence is important, but not sufficient for the effective support and introduction of new technology.[4]

Perhaps more importantly, similar ideas seem to prevail for the developing countries. Where once it was thought that the establishment of a scientific community of a critical size in a developing country would solve most of its problems in time, it is now obvious that this is not true. For example, India now ranks third in the world in terms of the number of scientifically and technically trained personnel—but it is more than ready to admit that it has yet to find ways of using most of them effectively. As S. Bhagavantam, President of the Indian Institute of Science, recently put it:

instead of hastily concluding that we should increase the size of the scientific community in each of the developing countries, the issue should be looked at a little more critically . . . while a large-sized scientific community is at best a necessary requirement for development, it certainly cannot be a sufficient requirement . . . it is not the size of the community that really matters, but the quality.[5]

F. F. Papa Blanco takes the point a stage further: 'Highly qualified people in a country are potentially a resource', he writes, 'but if they are not properly integrated in the development process, they may well become a social and economic load.'[6] Or, indeed, part of the brain drain.

Furthermore, the whole idea of fundamental research leading to technology and that in turn leading to development has come to be questioned. Alexander King, commenting on the fact that so much of world research and development is undertaken in the industrialized countries, claims that it is 'difficult to be sure to what extent their economic success is due to this and how much their high research intensity has arisen because they could afford it'.[7] Herbert Fusfeld is much more direct:

in the broad sense, basic research *follows* applied research, and research and development *follow* industry. That is, once an industrial base is established, money and manpower can be devoted to R and D because the system for using the results is in place. Once a major development is well underway, there is increased justification and pressure for basic research and a sharper definition of the particular areas of basic research where new knowledge would be most helpful.[8]

If this is true, we have a real chicken-and-egg problem. Research and development are always presumed to be needed to stimulate economic growth; but if research and development more normally follow in the wake of industry and progress, the lever which it was thought that science had on development looks a great deal less effective. It is extraordinary that after three decades of fairly intensive work on the problems of global development, the answers to such basic questions should be in any doubt at all.

The issue affects the developing countries most acutely. They are currently engaged in a determined effort to build up their own scientific communities, arguing that such communities are essential if the Third World is not to be permanently dependent on the industrialized nations for its science and technology. The implicit argument is that, when these new scientific communities reach some critical level of size and/or excellence, the developing countries will begin to be able to generate their own technologies. But if Fusfeld is right, the developing countries might do better to concentrate instead on building up an industrial base. This base could later generate the wealth needed to fund research and simultaneously define the areas in most pressing need of research.

However, this argument must not be taken too literally. It is included here as an illustration of our ignorance of the way science, technology, and industry interact. At most, Fusfeld's view indicates that the view that 'science transfer to developing countries should always precede technology transfer'[9] may be somewhat hasty. As Klaus Gottstein writes: 'The answer is probably that there should be a parallel transfer of science, applied research, technology and industry in a well co-ordinated, balanced manner. Science cannot flourish in isolation.'[10]

In May 1979 Unesco organized a meeting of scientists in preparation for the UN Conference on Science and Technology for Development. Among those present was Gerald Holton, Mallinckrodt Professor of Physics and Professor of the History of Science at Harvard University. He made a special plea to Unesco:

apart from certain specific victories, we are still very ignorant about fundamental questions about the dynamics of science, technology and development. Any historian of science and technology will tell you that we are only beginning to understand the relation between science and technology (namely, which 'pulls' which). We don't know why in almost all countries the level of scientific activity is linearly proportional to the GNP, but in a few countries it is markedly greater . . . There is now a whole profession devoting itself to such studies—sociologists and historians of science and technology, some economists, some science policy scholars. I therefore strongly urge that Unesco involves the attention and dedication of an international group of such experts, specifically to develop models for understanding science, technology and development, without which all our goodwill and our intentions will be much less effective.[11]

This book owes its existence to this, and other similar advice which has been given to Unesco. The Advisory Panel on Science, Technology, and Society was set up precisely to help clarify the issues raised by Holton and others.

The fusion of science and technology

One of the ways in which our understanding of these things certainly has changed concerns our ideas about the roles of basic research, applied research, and development. During the 1970s it became clear that the old distinctions between science and technology had become blurred—basic research led to application so reliably, and problems of technological development provoked scientific enquiry to such an extent, that it became virtually impossible to distinguish between basic and other kinds of research simply by the results produced.

Science and technology thus seemed well on the way to forming an indissoluble whole, far different from the convenient marriage of two distinct partners a century previously. Such a transformation, of course, fulfils the Marxist view of the social role of science and technology. As V. N. Turchenko has written: 'The scientific and technological revolution constitutes a radical qualitative transformation of the productive forces achieved as a result of the complete merging of the scientific and technological revolutions and the transformation of science into a direct productive force.'[12]

The fusion of science with technology is not, however, always viewed so positively. The occasion when it first occurred in the field of molecular biology was in 1977 when it was first shown that a protein, in this case the human hormone somatostatin, could be synthesized by a human gene working inside a bacterium. The occasion was a double triumph: scientifically, it had been proved that a human gene could function in this way; and, technologically, it had been shown to be possible to manufacture protein using a new technique. In this case, the same action was both science and technology. Inevitably, confusion arose. Normally such an achievement would have been first published in an academic journal. As it was, the announcement was first made before a Senate Subcommittee by the President of the US National Academy of Sciences.

Some time later scientists succeeded in making interferon using similar techniques. This time the announcement was made at a press conference and shrewd commentators were quick to point out the subsequent rise in value of some shares with which the scientists concerned were associated. This is an example of the complete fusion of science with technology.

This mutual interpenetration of science and technology has a number of social implications: on the one hand, it helps protect science from the charge of irrelevancy, so common during the 1970s; and on the other it leads, as we saw in the previous chapter, to the charge of 'inevitable technology', thrusting a perhaps unwanted sense of responsibility on to scientists for the implications of their work. It is, of course, absurd, as Sir Peter Medawar has put it, to 'blame the weapon for the crime'.[13] On the other hand, it is now a well-established principle of international law that ignorance of or indifference to ends is no longer a permissible defence for those engaged in work on the means. And to most people it does seem entirely reasonable that if science is now held to be relevant because of an intimate and inevitable relationship with technology, it should take the rough with the smooth—credit for the good things in life must often be accompanied by at least some responsibility for the bad.

However, if it is true that science is impossible to distinguish from technology simply by reference to results, there are certainly other, more sophisticated indicators which reveal differences. For instance, it can be argued that scientists work towards discovery and technologists towards invention; or that, as Derek de Solla Price put it, scientists want to write and technologists to read; or even, as Jean-Jacques Salomon has argued, that the only real differences lie in the motivation of the individual scientists concerned or the nature of the institutions in which they work. But none of these definitions really helps to illuminate the workings of the system Holton was referring to.

Can research be targeted?

Increasingly, however, the world of science is now being viewed through different spectacles. More and more often the view is taken that in fact all science has an aim—an aim which may be to develop a better process for converting sugar to alcohol, or to improve understanding of the migration of some species of birds, or to find a cheaper way of colouring beads produced by an indigenous technology. A scientific effort is therefore directed at these goals—and we call the process itself 'targeted research' and the outcome 'directed knowledge'.

It should be stressed that this is more than just a change in nomenclature: it reflects a radical change in social attitudes towards science which has been going on for twenty years or more. Science is left to follow its own internal logic increasingly less; and more and more often it is directed towards the solution of some specific problem, a problem which may be defined either in terms of practice or of theory—the distinction no longer being so important.

The important thing that has changed here, of course, is the process by which science advances. The logic of direction used to come, to use Paul Goodman's famous phrase, from a 'wandering dialogue with the unknown'. The dialogue continues, but the wandering is sharply curtailed, the preferred directions carefully indicated.

It remains to be seen whether the cost of making science relevant is that of replacing a wandering dialogue with the unknown by what many might call a route march towards the pragmatic. Certainly there are plenty who seem to think so. One of Unesco's advisers on science, technology, and society, Ibrahim Helmi Abdel Rahman, writes:

Scientific studies are explorative and contemplative and scientists must be left free to widen the frontiers of human knowledge, and open new paths to the unknown, whether in the exploration of the innermost structure of matter or the widest reaches of the space–time continuum. The theories of science are logical structures invented by science as temporary 'scaffolds' to build knowledge. They are dispensable and may eventually be replaced by others.[14]

Clearly, if scientific knowledge consists of a series of scaffolds from which the next advances are built, targeted research makes little sense; advances can come only where the proper scaffolding already exists. There is plenty of experience to indicate that attempts to focus science on problems which are scientifically immature are usually unsuccessful. Of the situation in the United States, Liebe F. Cavalieri writes: 'The National Cancer Act of 1971 was conceived as a crash program, similar in principle to a moon landing. But basic scientific knowledge was in no way adequate for a technological approach to the cure for cancer. For a number of years, funds were lavished on mission-oriented programs that have not (and could not have) lived up to the expectations of Congress and the public.'[15]

The tension that exists between the dictates of the internal logic of science and the needs of society is the major problem of national science policy-making. Suffice to say here that science is now rarely regarded as completely autonomous, and the *laissez-faire* model of science policy associated with M. Polanyi[16] is currently unfashionable. This does not, of course, deny that science does indeed possess its own internal logic—the major problem of science policy-makers is to find the means of reconciling this logic with the often compelling needs of society. These ideas have recently been examined in detail in the Unesco publication *Societal Utilization of Scientific and Technological Research*,[17] in which 'knowledge push' science policy models are contrasted and compared with 'demand pull' models.

The fact that a certain field may not be scientifically ripe for a research-based onslaught is actually only one of the problems of targeted research. Even greater problems surround the issue of how to direct science at social issues. David Collingridge, who has made a special study of the social control of technology, sums up what has been a major technical dilemma for more than two decades:

our understanding of the physical and biological world in which we live is extremely deep, and provides us with the means for the production of all kinds of technical marvels; but our appreciation of how these marvels affect society is parlous. Ask technologists to build gadgets which explode with enormous power or to get men to the moon, and success can be expected . . . But ask them to get food for the poor; to develop transport systems for the journeys which people want; to provide machines which will work efficiently without alienating the men who work them; to provide security from war, liberation from mental stress, or anything else where the technological hardware can fulfil its function only through interaction with people and their societies, and success is far from guaranteed.[18]

It is paradoxical that this situation should still prevail when today science is so often accused of masquerading as technology. This is a double problem, and the crux of the science/society dilemma. On the one hand, it is curiously difficult to direct successful research programmes at real social problems —as the Green Revolution showed, we can find ways of making more food but not apparently of feeding more people. And on the other hand, it is almost equally difficult to prevent research giving rise to new technologies, whether or not there is a need for them. As Cavalieri concludes:

We are no longer in an era when practical applications of scientific research are unforeseeable and the human consequences unknown. Most of the science practised today has at least a speculative relationship to potential technology, and even when that is not true, we know enough about the relationship of science to technology and technology to society to know that caution is advisable. It has become the moral responsibility of scientists to consider the social implications of what they are doing—partly because they are inevitably the first to sense the approach of new technological capacities, and partly because there is no one else ready to take the burden of responsibility from them.[19]

We shall return to this issue of the social control of both science and technology later in this book—Part III deals with the choices we have to make in the future conduct of our scientific and technical affairs. For the moment, however, we concentrate on the relationship of science and technology to development, on the distribution of the world scientific effort, and on the relevancy of its major goals to the pressing problems of the end of the twentieth century.

3 Technology for development

For scientists in the Third World, the application of science to the task of overcoming underdevelopment represents one of the greatest moral and intellectual challenges of history.

Amilcar Herrera, 1971

Even twenty years ago, the idea that scientists from the Third World could themselves overcome the problems of development would have been a novelty. At that time, less than two decades after the Second World War, the concept of development was still relatively young—it dates, in fact, only from the end of the Second World War when politicians from the victorious nations tried to make reparations to other countries which had suffered heavily in the conflict. Those reparations took two principal forms: food, to alleviate immediate problems, and arms, to ensure that 'friendly' governments managed to keep their political opponents under control. To these two ingredients was later added financial aid in the hope that a catalyst seeded at the right time would boost the economy of a developing country just sufficiently to allow it to 'take off' of its own accord.

In all this, the role of science and technology became progressively more important—though it was not until much later that the idea of endogenous science came to the fore. This was what has been aptly called the 'euphoric' period for science and technology. Science and technology, it was thought, would eventually enable the developing countries to enjoy all the riches of the Earth: the deserts would be made to bloom, human disease would be finally conquered, and the cities of the Third World, if not actually paved with gold, would certainly be filled with prosperity. In the industrialized world confidence abounded. Thus in 1967, when the first dark clouds were beginning to appear on the horizon, the Dean of the Massachusetts Institute of Technology's College of Engineering was able to boast: 'I doubt if there is such a thing as an urban crisis, but if there were MIT would lick it in the same way we handled the Second World War.'[1]

During this period, too, there was a great deal of concern with the 'technology gap'. First the United States and later Japan began to show impressive rates of economic growth—considerably higher than those of most countries in Europe or anywhere else. This was generally assumed to be because the rate of technical innovation in those countries was higher than elsewhere. The key to growth was thought to be innovation; and the key to innovation was thought to be research.

However, this somewhat facile view of progress was not to last long. In the same year as MIT's Dean was offering to control urban chaos, Edward Denison published a major study of the factors which affected economic performance.[2] This showed that rate of innovation could not be used as a single explanation of why some countries' economies grew faster than others. Denison listed more than thirty other variables, including such things as health and use of capital, which significantly affected rates of economic growth. Denison's analysis was thoughtful; and it and the many studies to which it later gave rise are essential reading for anyone who still believes that there exist magical solutions to development problems.

Politics and technology

In a sense, this marked the end of the euphoric period for science and technology. During the 1970s, the complexity of the development process and the factors affecting it were studied in increasing detail. The brash optimism of development experts in the developed countries was replaced by a more thoughtful approach and, it must be admitted, by a certain cynicism—for the first time doubts began to be expressed as to the feasibility of rapid world development. At the same time, the wishes of some of the developing countries began to be expressed a great deal more forcibly. The notion of the New International Economic Order which was formulated in 1974–5 marks a completely new conception of development, and of the problems which cause lack of development. One unfortunate result has been a worsening of relations between the rich countries and the poor ones.

As far as science and technology were concerned, one of the key issues which underlay many of the debates of the 1970s was whether development was being held up through lack of technology or through lack of political action. In fact, of course, it was not a simple question of choosing between two absolutes but a much more complicated issue of deciding where the priorities lay. Undoubtedly, during the 1970s the pendulum swung far towards the political explanation: for example, it became increasingly fashionable to explain the world food shortage not in terms of insufficient production but in terms of poor distribution systems and unfair marketing regulations which produced artificial prices.

According to this view political reform, coupled perhaps with a wiser use of existing and new technology, was what was required to alleviate the problem. In its five-year outlook published in 1979, the US National Academy of Sciences is quite specific about the experience of the 1970s: 'We have learned from sad experience that applying science and technology to political problems does not solve the problems.'[3]

One of the ways in which this lesson was finally learned was during the 'Green Revolution'—an event of such importance in the science/society debate that it is worth examining in some detail. The background, of course, was the discovery of new high-yield strains of various domestic crops—notably maize, wheat, and rice—which promised to increase food production in tropical areas—notably south-east Asia—very considerably. And when they were introduced in the late 1960s, they did indeed increase production. But there were other effects as well.

The new strains depended for their effects on a high rate of fertilizer application and on excellent irrigation. The seeds were hybrids, and so new seed could no longer be grown by the farmer himself but had to be purchased annually at increased cost. The increase in production that was made possible also required that storage capacities were greatly increased, new roads built to deliver the essential fertilizers, and better means introduced to control rodents and other pests that feed on the product, which often had to be stored for longer periods.

Among the unexpected, or at least insufficiently studied, results was the fact that the larger farmers tended to increase their holdings at the expense of the smaller farmers, who, forced out of business, then swelled the ranks of the urban unemployed. The latter tended to benefit little from the new strains, for they could not afford to buy the extra food that resulted from them. Indeed, very often that food was not available for purchase; much of the increased production had to be pledged for export in order to pay for the new fertilizer plants, roads, and storage facilities. Thus the developing countries became tied even more closely to the world agricultural economy, and the marginal producers continued to suffer from its fluctuations.

The story of the Green Revolution seems to provide a case history to illustrate almost everyone's point of view. Technocrats frequently claim that it provides a splendid example of what science could do for the world if only political obstacles did not get in the way. The advocates of appropriate technology use it as an illustration of the myriad side-effects which can accompany the casual introduction of foreign technology into a rural economy. But it also illustrates the technology/politics dilemma.

Faced with a problem, does one turn first to a political or a technical solution? Or does it depend on the nature of the problem? Clearly, the Green Revolution was a technical solution and it did not achieve all the objectives which might have been wished for. On the other hand, a political solution—involving land reform, economic changes, and improved support for the deprived—might well have worked better but would undoubtedly have been far more difficult to implement. Indeed, it is the very complexity of political solutions which makes the apparent simplicity of technical alternatives so attractive. This simplicity can, however, be deceptive. Essentially, this choice is a familiar example of lateral thinking: if you can't agree on how to divide the cake fairly, then make the cake bigger.

Classifying the problem

This uneasy relationship between science and politics has caused many people to try to categorize development problems into two groups—those amenable to technical solution and those amenable to political solution (we should also note here the existence of a third, far less commonly discussed category of problem— problems which have no solution at all[4]). The following scheme was drawn up in a paper prepared for a meeting of the Unesco Advisory Panel on Science, Technology, and Society in September 1981.[5]

Problems amenable to science/technology

- hunger and malnutrition
- illness and health
- deterioration of the biosphere
- natural disasters
- illiteracy
- lack of information for technical or linguistic reasons
- poor rural and urban management
- absolute poverty
- the energy crisis
- depletion of resources, including water
- loss of cultural identity
- alienation.

Problems not amenable to science/technology

- the arms race and war
- the rise in violence
- erosion of human rights
- lack of information for commercial, legal, educational, or psychological reasons
- unemployment and inflation
- disproportion of income among citizens
- dislocation of international trade.

A close inspection of all such lists reveals that the problem of classification is not as simple as might at first sight appear. For example, although illness may be cured and health improved, it seems unlikely that the process can continue indefinitely: the elimination of human disease is almost certainly a category III (no solution) problem. Similarly, natural disasters are never likely to be eliminated (though their impact may well be lessened). On the other hand, issues such as alienation, loss of cultural identity, erosion of human rights, and disproportion of income may all arise or be worsened as a result of scientific/technical advance, even though they are listed as problems not amenable to science/technology—in other words, their relationship to science and technology may be far more intimate than the list implies. Similarly, as the Green Revolution shows, although a problem may be classified as purely technical, the social, cultural, and political issues must all be considered very carefully if a technical success is to have any meaning in the long run.

The Unesco paper is careful to point out that some of the category I problems really stem from a one-dimensional or alienating concept of development. The category II problems, it claims, may stem *inter alia* 'from an incorrect use of S and T and can only be solved by a change in behaviour, attitude and values, and not through science'.[6]

However, many of the category I problems can also be cured by a 'change in behaviour, attitude and values'. For example, the state of the biosphere can certainly be improved by the use of the right technologies. But it is also easy to envisage improvements resulting from an increased sense of personal responsibility towards the environment or from legislation preventing deliberate pollution. Similarly, energy problems can be approached in one of two distinct ways—find more energy (science and technology required) or use less energy (science and technology not necessarily required, although an understanding of the principles involved is).

Closer analysis therefore suggests that such problems cannot be simply classified as amenable to science/technology or not amenable to science/technology. Most political scientists argue that the decision on how to categorize a problem is itself a political decision, and a major one. Such a decision will reflect the basic characteristics of the society concerned, and determine its attitude towards the world at large.

Science and scarcity

A Canadian political scientist, Michael W. Jackson, has made an elegant analysis of this science/politics dichotomy.[7] According to Jackson, 'To emphasize science as the solution to the problems of public policy is to diagnose scarcity as cause and to prognosticate wealth as cure.' A society which does so is then likely to assume an increasingly technocratic stance—for example, it may view conflict as caused primarily by a battle for scarce resources rather than as an expression of ideological difference. It is likely to champion individual rights rather than group responsibilities. And as scientific criteria are applied more often in preference to political and ideological criteria, 'ideology is surpassed by science almost as Marx saw the State ultimately wither. Our ambitions are given their opportunity by the orderly processes of science, not by the disorderly processes of politics.'

While all this may seem academic, in fact it provides some real clues as to how the world works. For example, Jackson's theory explains why a nation suffering from severe internal conflicts— facing, for example, racial, tribal, or religious schism—is likely to look for technocratic solutions to its problems simply because its political choices are so difficult.

At the end of the day, Jackson warns, two dangerous myths must be checked. The first is the American tenet that with enough science any nation can reach an ideal state of affairs; the second is the contemporary European idea that with enough good policies any nation can become as powerful and rich as the United States. 'Neither science nor policy, separately or together, can produce richness and well-being if other, essential resources are not present in the land or among its people', Jackson claims.

He goes on to provide a careful warning against the over-use of technology in situations where it is inappropriate:

Technology and the technocratic outlook are somewhat like hammers and the 'law of the hammer': give a child a hammer and it will discover that everything needs pounding. There is a bit of the child in us all when we are armed with a wondrous tool of technology, and we need to guard against the temptation to bang everything with it. We need the hammer, but not the impulse to over-use it.[8]

The moral of this for nations seeking rapid development seems clear. An analysis must be made of which of the nation's problems are really caused by scarcity, and which are apparently caused by scarcity but in fact result from policies imposed elsewhere. According to Jackson, the former may respond to a scientific/ technological effort; the latter never will.

Changing the priorities

That real scarcity exists in all developing countries is not, of course, in dispute. However, two major problems face any developing nation wishing to use science and technology for development.

First, the balance of research is very heavily weighted towards the industrialized world—not only is most of the research carried out there but most of the people qualified to do it (the scientists and engineers) also live there. In 1974, of the 2,978,204 scientists and engineers engaged in research and development throughout the world, 93.9 per cent worked in the developed countries, 5.8 per cent in the developing countries and only 0.3 per cent in the least-developed countries.[9]

According to the same source, world expenditure at that time on research and development amounted to $101,785m. (this figure had increased to more than $150,000m. by the beginning of the 1980s). Of this, $99,158m. (97.4 per cent) were spent in developed countries, $2,544m. (2.5 per cent) in developing countries, and only $83m. (0.08 per cent) in the least-developed countries.

The maldistribution of research and development is even more dramatically illustrated in terms of the money spent on research and development per head of population. On this basis, in 1979 the United States spent $200 on research and development per head of population. Expenditure in the Latin American countries was typically less than $5 per head and in most countries in Africa and Asia it was less than $1.

To make matters worse, the priorities within the developed countries are of little relevance to the Third World. In many of them as much as 70 per cent of all research and development is directed at defence, space, or nuclear energy. In others, the problem is even worse. During 1982–3, for example, the United Kingdom government planned to spend £2,750m. on research— £1,750m. of it, or 63.6 per cent, on defence research, and only £1,000m., 36.4 per cent, on civil research. Globally, according to Colin Norman, 'Military R and D alone accounts for more financial and intellectual resources than are devoted to R and D on health, food production, energy and environmental protection combined.'[10]

It is difficult to find accurate information on the global distribution of research priorities. Norman has published his own assessment[11] but warns that the figures (Table A) should be regarded only as a rough guide.

One other aspect of global research priorities also needs further investigation. In many developing countries, there is a preponderance of highly theoretical research, unrelated to the needs of the country. Of course, a balance between fundamental and targeted research does need to be maintained but the scientific

Table A: Global research and development budget 1980

Programme	Percentage share
military	24
basic research	15
space	8
energy	8
health	7
information processing	5
transport	5
pollution control	5
agriculture	3
other	20

community in a developing country is often tempted to stress theoretical studies simply because in that way it is more easily accepted as part of the international scientific community. This issue is dealt with in more detail in Chapter 8 ('The scientific community').

Even where research is directed at pressing social or economic problems, the necessary links may not exist between the research and the production sectors to turn the results into concrete action. In developing countries, research of this kind, and highly theoretical research, is thus likely to prove of more benefit to the industrialized world than it is to the countries in greater need. As F. F. Papa Blanco writes, 'Basic research undertaken under inappropriate conditions, or where other factors make further development impossible, becomes a subsidy that the less developed countries give the developed world.'[12]

There are other ways in which the new knowledge generated by science and technology appears not to help the developing countries. Indeed, under the present situation there is a curious anomaly by which scientific advance may actually impede progress. It is usual to think of the pool of scientific knowledge generated by the world's scientists as continuously expanding— new knowledge is added but the old knowledge never forgotten. In fact, there is some evidence that things do not actually work out like this in practice. Instead of a constantly expanding store of knowledge from which all can draw, what we see, according to Hans Singer, is:

much more comparable to a flow than to simple accumulation. New science and technology are created at one end, but displace science and technology that existed previously. Being located in the rich countries with their dynamic search for new knowledge and their vastly and increasingly different requirements and priorities, it is not surprising that the knowledge displaced or submerged at one end may well be more useful to the developing countries than the new knowledge added at the other end. It is therefore by no means clear, from the viewpoint of the poorer countries, that there is in fact an accumulation of knowledge in the relevant sense.[13]

In other words, the developing countries not only fail to benefit from the new knowledge created elsewhere, but that new knowledge in practice destroys—or more precisely makes un-available—older technologies which could have been of use. In much the same way consumer products quickly become obsolete because spare parts for older models soon become unobtainable when new models replace the old.

This is not a new situation, and several attempts have already been made to deal with it. After the United Nations Conference on the Application of Science and Technology for the Benefit of the Less Developed Nations (UNCSAT, Geneva, 1963), a new UN body—the Advisory Committee on the Application of Science and Technology to Development (ACAST)—was formed. During the 1960s it prepared a World Plan of Action[14] for science and technology, published in 1971, which was meant to provide the scientific and technical input for the UN Second Development Decade. (By 1981, after the UN Conference on Science and Technology for Development, held in Vienna in 1979, ACAST had been transformed into the new Intergovernmental Committee on Science and Technology for Development.) Among ACAST's many proposals were that:

- by 1980 developing countries should spend at least one per cent of their GNP on research, development, and the provision of scientific services
- developed countries should spend 0.5 per cent of their GNP on science and technology aid to developing countries
- developed countries should devote 5 per cent of their non-military research and development to the needs of the developing countries.

At the time, these measures would have had the effect of directing about $8,500m. worth of research towards the problems of development. In the event, nothing like this amount of money was found and the World Plan of Action, according to ACAST's Chairman, W. K. Chagula, 'remains a document destined to gather dust on the shelves of libraries and archives'.[15] By the late 1970s, 'barely one per cent of spending on research and development in the North was specifically concerned with the problems of the South; whereas 51 per cent was devoted to defence, atomic and space research. The importance of disarming as a possible means of promoting development is nowhere more evident than in the field of research.'[16]

The problems of technology transfer

Research is, of course, only one means of acquiring the technologies that may be needed for development—and it is the one which operates on the longest time-scale. In the short term, the developing countries are forced to obtain their technologies by other means. Since the early 1970s a great deal of thought has gone into examining the future role of technology in development. In particular, the problems of technology transfer have been exhaustively analysed.

By the end of the 1970s the direct costs of technology transfer in the developing countries were estimated at $9000m. to $10,000m. a year. Indirect costs, such as those accruing from price mark-ups and the over-pricing of intermediate imports, were considerably greater. According to Surendra J. Patel, chief of UNCTAD's Transfer of Technology Division, they probably amounted to $30,000m. to $50,000m. a year.[17] By comparison, the developing countries were spending only $2000m. a year on their own research and development at that time.

Complaints about the inequity of this situation dominated debates during the 1970s—the complaints concerning not only the cost of the operation but also the fact that the developed countries subjected the less developed to technological dependence or domination. As Francisco R. Sagasti, a Peruvian expert on technological development, put it:

To dramatize the situation from the perspective of the Third World, a few hundred people in the highly industrialized nations now make decisions on who is going to get which part of the new technologies at the world level, and under what conditions . . . we are witnessing a shift towards the use of access to modern technology as the main vehicle for exerting control over the productive activities of Third World countries, showing once more the truth of the Baconian precept 'knowledge is power'.[18]

Within the United Nations the Group of 77—representing the developing countries—began to call for a New International

Technological Order which they argued was needed to comple-
ment the New International Economic Order. And for more than
a decade the United Nations Conference on the Code of Conduct
for Technology Transfer has been busy trying to hammer out
details of how licensing arrangements can be improved, what
kind of role the transnational companies should play in future
development, and how the developing countries can generally be
given a better deal in the international technology fair. Although
by the early 1980s, this action had still not produced an agreed
Code acceded to by all UN member-states, there are prospects for
success (see Chapter 11, 'The transnational enterprise').

This debate reminded many of the debate over the technology
gap between the United States and Europe in the 1960s. As an
OECD report put it: 'Then the cry was for a "Marshall Plan for
science"; now the call is formulated as "free access to technology, the common heritage of mankind". These demands are
variations on the same theme—it is only the style that has
changed.'[19]

The OECD thought that the technological advantage of the
West might not be of long duration. It noted that only three years
after a major meeting on the technology gap in the 1960s, US
exports of high technology products began to fall, and imports to
rise. Japan was on the way up, and some of Europe was close
behind her. Soon afterwards concern about the technology gap
began to wane and the issue has not reappeared.

. . . and of cultural transfer

It would be wishful thinking to imagine that a similarly neat
solution could await the problem of technology transfer. A fairer
deal for the developing countries will have to be obtained, and it
will have to be obtained through international legislation.
However, the current inequities of the situation are not the only
problem.

According to a major report prepared by the International
Federation of Institutes of Advanced Study (IFIAS) and edited
by its chairman, Alexander King:

The real difficulties of transfer do not lie so much in the inadequacies of the present system as such, the machinations of the transnational corporations, or the inequities in the conditions of transfer, but in a failure on the part of both donors and receivers to appreciate that successful transfer is an exceedingly complicated socio-economic process with many facets and social conditions beyond the mere introduction of packaged processes and know-how. Above all it necessitates the existence in each country of an indigenous capacity for science, technology and industry, if imported processes are to be assimilated, take root and spread.[20]

By this time, of course, the message of those who had been urging that technology should be made appropriate to the conditions in which it was expected to work had finally struck home. The inappropriateness of simply transferring Western technology—with all its inherent social and cultural values—to a foreign country is now almost universally recognized. Even the research director of a transnational corporation can now say:

Western science, technology and know-how can be important to developing nations, but not as isolated packages. There must not only be a receiver, but a receiving system with which the transferred package must be compatible. We sometimes talk as if we are throwing a ball to a waiting catcher when we should be considering the complexities of an organ transplant.[21]

This is a point that needs examining in more detail. If technology transfer involves something of the complexity of an organ transplant it follows that the receiver as well as the donor should be technically literate. There is a need, in other words, even for countries which rely entirely on technology transfer to build up their own science and technology capabilities. However, as Joao Frank da Costa—the Brazilian diplomat who was Secretary-General to UNCSTD—has put it: 'Development must neither depend on the importation of outmoded technologies from developed countries nor even advanced technology developed somewhere else. Only the mastery of world scientific knowledge, including the knowledge which for economic reasons has never been applied in developed countries, can assure an original, creative technological development.'[22]

The main reason for buying technology, rather than developing it, is surely because either it is cheaper or the highly qualified manpower required is not available. If, however, imported technology also requires the presence of an endogenous capacity for science then some, at least, of the motivation for importing the technology is lost.

How technology can be acquired has become a widely studied subject during the past decade. The old dichotomy of whether to buy or try to generate technology has been replaced by a spectrum of choices. Essentially, there are three possibilities:

- assess and improve indigenous technologies
- assess, adopt, or adapt foreign technologies
- develop new technologies.

The first of these is a more or less new ingredient—and one which stems not so much from technical necessity as from social and cultural requirements. As Unesco's Director-General, Amadou-Mahtar M'Bow, has put it:

it is advisable to consider assigning a significant role to endeavours to reinstate technical traditions based on age-old skills which have long been neglected in the name of modernity. Because they tend to be regarded as a way of reasserting the cultural characteristics of a society, they can be an excellent means of winning the confidence of the whole population, and not just of an elite, so as to start a general movement towards a form of development benefiting widely from the participation and initiative of every individual. This approach is likely to prove successful in stimulating endogenous creativity.[23]

However, the development of both indigenous and appropriate technologies requires qualified personnel, familiar not only with the scientific 'state of the art' of their field but also with the social and cultural requirements of their country. In the past few years it has also become clear that importing and adapting foreign technologies, as well as inventing new ones, is no less technically demanding. In fact, whichever route is chosen technical expertise is required. There is no longer any escaping the need for a national capability in science and technology. It involves the careful building up of institutions for education and research, and the fostering of a cultural climate in which both can flourish. The decision to embark on such a course is one of the most important political acts facing any developing country.

4 Development for what?

Science and technology *cannot* be applied to development. Science and technology are an essential part of development. One does not apply one's lungs to respiration, nor one's heart to the circulation of blood nor one's legs to walking. If we regard science and technology as a crutch, it will at best provide a halting gait. If we regard them as a transplanted heart, it will sooner or later be rejected by the receiver.

Thus did Dr H. B. G. Casimir, a past President of the Royal Netherlands Academy of Arts and Sciences, open his address to a UN colloquium on science, technology, and society held in 1979.[1] In claiming that science cannot be used for development, Casimir was not trying to denigrate science; but he was, by implication, criticizing the development goals of the past three decades.

What is development for? Obviously, there are many negative answers: to do away with malnutrition, poverty, disease, urban slums, rural stagnation . . . the list is long, and applies both to developing and to some of the more developed countries. But positive interpretations of development are harder to come by—what are the ultimate goals, the desired ends?

In the last analysis development must be viewed as a positive process, aimed at fulfilling mankind's highest aspirations. And if one of these be to understand the world in which we live, the practice of science must surely be included in the agenda for development. 'Science for development' is one thing which may or may not work; 'science as development' is quite another.

Progress from within

Each decade seems to generate its own catchphrases. 'The gap' dominated our thoughts for nearly a decade; 'the crisis' (of the environment) came soon afterwards; and the problems of technology transfer seemed to dominate the 1970s. Now, in the 1980s, we are firmly in the decade of the endogenous development of science and technology—a catch-phrase which itself manages to include many other ideas within it. It hints at self-sufficiency; it suggests that science, as well as being used for development, should also be considered as part of it; and it looks forward to a new international technological order when the developing countries can stand on an equal technological footing with the developed world.

It is always hard to pin-point the origin of ideas. And quite where the concept of endogenous science and technology came from is unclear. Its importance grew as did the idea for a new international economic order (NIEO) during the 1970s. Indeed, the programme for the NIEO, voted on in the UN General Assembly in 1975, puts the 'establishment, strengthening and development of the scientific and technological infrastructure of developing countries' first among the science and technology objectives.[2] Subsequently, the concept became an *idée fixe* during the preparations for the United Nations Conference on Science and Technology for Development (Vienna 1979). And on 31 August 1979 the delegates from 142 states adopted the Vienna Programme of Action, aimed at:

- strengthening the science and technology capacities of the developing countries
- restructuring the existing pattern of international scientific and technological relations
- strengthening the role of the United Nations system in the field of science and technology.

The Introduction to the Programme of Action states that 'the primary responsibility for the development of developing countries rests upon these countries themselves . . . The full recognition of the necessity for all countries to rely on their own endogenous scientific and technological capacities has characterized the preparatory activities for the Conference. Such self-reliance does not mean autarky but the ability, in essence, to take and implement autonomous decisions for the solution of national problems, and the strengthening of national independence.'[3]

The recommendations went on to outline many sensible but hardly revolutionary changes which would help developing countries expand their scientific activities and enable them to get a fairer share of world technology. For example, at the national level all developing countries were urged to do such things as set up a body responsible for science and technology policy-making, integrate research priorities with development plans, develop a capacity to 'unpack' imported technologies, and improve working conditions for the local scientific community with a view to solving the brain-drain problem.

Two other sets of recommendations dealt with regional and international issues. Regional actions were seen as part of a programme to encourage the idea of 'collective self-reliance' among the developing countries. Here specific objectives included the making of regional manpower inventories, developing regional projects to explore and use natural resources, and setting up regional centres for the transfer and development of technology. Finally, at the international level, the developed countries were urged to devote more of their research to problems relevant to developing countries, help the latter towards the endogenous development of their own scientific resources and to do something concrete about halting the brain drain. Throughout, great emphasis was placed on restructuring the world scientific and technological information systems so that the developing countries could gain their fair share of the knowledge already developed.

For many of the developing countries, these recommendations were but a watered-down version of a set of far more radical demands made by the Group of 77—the informal body of some 120 countries which now collectively represents the interests of developing nations in the UN system. After much debate, these demands were not agreed to; and in the end the really noteworthy things to come out of Vienna were a new and stronger UN committee to handle science and technology for development (the Intergovernmental Committee on Science and Technology for Development), a target of \$250m. for 1980–1 for an Interim Fund for science and technology for development (in place of the

$2,000m. by 1985 and $4,000m. by 1990 originally requested) and the promise to investigate long-term mechanisms of financing science and technology for development more effectively. In the event, even the more modest financial target, to have been provided by voluntary contributions, was not achieved; only about $40m. were raised for 1980–1.

These, at least, were the recognizable products of the conference. Meetings of this kind, however, often produce results which are not directly expressed in their formal outcome. And if any one notion could be said to have resulted from UNCSTD, it was that of the endogenous development of science and technology—a notion which, after UNCSTD, became the corner-stone of scientific development, accepted equally by members of both developing and developed countries.

Rewriting the idea of development

The significance of the idea of the endogenous development of science and technology relates to the ways in which ideas on development itself have changed over the past fifteen years or so. In the late 1960s, perhaps for the first time since the Industrial Revolution, thinkers in the Western world began to have serious doubts about the concept of economic development. Many of these doubts were sparked off by environmental considerations, but scores of other issues soon became involved. By 1979 a Latin American scientist, Amilcar O. Herrera, could write:

The idea of the Western approach as being practically the only model for progressive societies has undergone radical changes, and a new approach to development has started to emerge. Its distinctive element is that it concentrates on human beings; the well-being of individuals will not be a by-product of indiscriminate economic growth, but a specific target whose attainment will condition the whole social and economic organization of the country.[4]

In fact, this process of rethinking development was well under way by 1979. The 'basic needs' approach swept most development institutions by storm in the late 1970s and emerged in the 1980s as an accepted gospel—but a gospel which was more often preached than practised. It had profound implications for the development of science and technology.

Basic needs are not so easy to define in practice as in theory. Of course, food, health, education, and housing are basic in most cultures and therefore not greatly disputed. But needs which may be defined as basic in some countries may not be in others. This is particularly true when the needs in question relate more to the human spirit than they do to material requirements—the value of knowledge, for example, varies widely in different cultures. To make a list of basic needs therefore implies choice, and the only means of making such a choice democratically is through public participation. Hence the basic-needs model of development becomes inevitably linked with public participation.

The new model of development also became linked with the idea of self-reliance, defined as 'the will to build up the capacity for autonomous decision-making and implementation in all aspects of the development process including science and technology. This approach to self-reliance is reflected internationally as opposition to all forms of dependency.'[5]

Basic needs, self-reliance, and public participation thus became the new keywords of development. How science and technology could be applied to development was thus altered, for development itself had altered. The technology to satisfy the new constraints was clearly the appropriate technology which had been invented in concept the decade before—but now appropriate technology was no longer a solution in search of a problem. The changing aims of development policy had elevated it to a position of some importance.

Clearly, these sweeping changes in development goals implied equally broad changes in research and development priorities. According to Hans Singer of the UN International Labour Organization, 'indigenous R and D . . . will have to give high priority to the needs of small farmers, rural and small-scale industries and the informal sector'.[6] In other words, such products as the *Tinkabi* tractor of Swaziland, the small-scale cement plants of India, the biogas digesters of China, and the rural housing techniques of Guatemala—to mention four of the best-known examples—should have become a prime focus, rather than a peripheral one, of research and development.

However, as Herrera commented, 'To devise a science policy to implement self-reliance is not an easy task. It is not simply to create technical solutions for certain problems; it is, above all, to incorporate in those solutions the specific characteristics of the society involved.'[7] Hence the technologies concerned must always be appropriate but they do not necessarily have to be endogenous. It is the process by which they are created or selected which must be endogenous.

Such a definition leaves a developing country free to develop its own technologies, rejuvenate its traditional practices, or import a foreign technology with or without adaptation. But the decision to do any of these things must always come from within the country. The endogenous approach also, of course, allows less-conventional attacks on the problem, such as the Chinese 'two legs' approach in which high technology exists alongside appropriate technology—an idea which initially seemed daring but which had become almost commonplace by the early 1980s

The twelve 'musts'

By the 1980s, too, the rapidly changing concept of development had clarified. Joao Frank da Costa, Secretary-General of UNCSTD, spelt out the new understanding in a list of twelve 'musts for development',[8] which for many people brought the blurred image of a rapidly changing scene into much sharper focus:

1. *Development must be total*—not just limited to economic factors but expanded to include all social and cultural considerations.

2. *Development must be original*—hence development styles must be diverse and antique concepts of 'the gap' can be dispensed with.

3. *Development must be self-determined*.

4. *Development must be self-generated*—but self-reliance will usually be achieved by horizontal co-operation with other developing countries or triangular co-operation involving developed countries as well.

5. *Development must be integrated*—for example, the industrial and agricultural sectors must be developed jointly, together with the system for education and training.

6. *Development must respect the integrity of the environment*—both natural and cultural.

7. *Development must be planned*—the free play of economic forces does not lead to an equitable diffusion of scientific and technological potential.

8. *Development must be directed towards a just and equitable social order*—all sectors of the population must benefit equally from the applications of science and technology.

9. *Development must be democratic*—the goals of society are not all scientific and technological; science and technology must not be allowed to assume control.

10. *Development must not insulate less-developed regions into 'reservations'*.

11. *Development must be innovative*—depending neither on the importation of advanced or outmoded technologies.

12. *Development planning must be based on a realistic definition of national needs*.

These twelve commandments of development, and their relationship to science and technology, made the situation a good deal clearer. To see how the developing countries themselves interpreted the problem, however, it is instructive to look at the Lagos Plan of Action[9] for the economic development of Africa during 1980–2000 which was adopted by the Organization of African Unity (OAU) in April 1980. The OAU tried to deal with both the problems and their causes, and then with their solutions.

In spite of all efforts, says the OAU, Africa has remained in a state of stagnation for twenty years and is more vulnerable than other regions to social and economic crises. It is the least developed of all the continents, with a combined gross domestic product of only 2.7 per cent of total world revenue, averaging $166 a year per head of population. And it includes twenty out of the thirty-one least-developed countries.

Why this should be is a paradox in view of the immense

resources of the continent. These include 97 per cent of world reserves of chromium, 85 per cent of platinum, 64 per cent of manganese, 25 per cent of uranium, and 13 per cent of copper, as well as 20 per cent of world potential hydropower, 70 per cent of world production of cocoa, 33 per cent of coffee, and 50 per cent of palm-oil.

The OAU goes on to argue that economic liberation can come only if these resources are put to proper use, and Africa stops depending so much on the export of raw materials. 'Africa', it says, 'must cultivate the virtue of self-sufficiency.'

In considering the problems of science and technology, the OAU laments the fact that most of its members still have inadequate national capacities and continue to depend on foreign technicians and experts. The reasons for this are listed as follows:

- the old international economic order continues to exist and influence the way technology is handled and controlled
- national mechanisms for developing endogenous technologies are inefficient
- imported foreign technology continues to have negative effects
- policies for science and technology are too imprecise
- lack of education and training in areas of science and technology relevant to development
- there is a dichotomy between urban and rural technologies
- lack of contact between the productive sector and research and development (which concentrates too much on areas of general interest and the acquisition of knowledge in non-essential areas)
- there is no mechanism for regulating technology transfer
- too little money is devoted to science and technology
- insufficient priority is given to science and technology.

The Lagos Plan of Action proposes a nine-point plan to deal with the situation. Essentially, it suggests that all African nations create a national science and technology centre for development, improve science and technology education and training, provide an institutional base capable of inventing, innovating, and adapting technologies for African development patterns, improve local production capacities in such industries as steel-making and aviation, develop cheap rural technologies, integrate research, development, and appropriate technologies with development goals for agriculture, industry, natural resources, energy, transport and communications, health, urban development, and the environment, and mobilize additional national funds for science and technology.

In addition, the Plan recommends the development of regional centres for science and technology and that Africa should seek and obtain substantial international funds for the development of its scientific and technological base.

From all this, it is clear that, in Africa at least, the main focus of attention during the next decade is on lessening technical dependence on the outside world and on creating endogenous capabilities in science and technology. However, while most development experts accept the wisdom of this general course of action, not all are convinced that the transition will be without problems.

Bootstrap critics

The idea of endogenous development, scientific or otherwise, has always suffered from one internal contradiction. Thus Victor L. Urquidi, President of the College of Mexico, writes:

It seems to me that 'endogenous development' boils down to the idea that things should start from the bottom up, in accordance with community or national desires and aims, however defined and formulated . . . [but] once the complexity of the Third World is admitted, together with its diversity of cultural conditions, it becomes less and less clear how endogenous development, at least in its narrower sense, can be carried out. Can a small community, or a small African country, pull itself up by its bootstraps? Is any rural community in a developing country, except for some tribal areas in Africa or North Borneo, immune to the information provided by the transistor radio? Are there any airtight cultures?[10]

What worries Urquidi and others like him is that the baby may be thrown out with the bath water. The move towards self-reliance and endogenous development is a reaction against domination and dependence; the dangers of that move are that the possibilities of extracting some of the 'goods' while still avoiding the 'bads' of domination may inevitably be lost. As Urquidi puts it, 'Refusal to consider the exogenous factors and possible advantages to be derived from them might only lead to no development at all, in effect to anti-development.'[11]

There are other problems associated with endogenous development. For example, is such a bootstrap philosophy compatible with ideas of international aid and solidarity? How much of the United Nations system is really needed if countries are actually going to manufacture their own development styles and directions? Indeed, is help from the outside even admissible with such a goal?

If this is a problem at the general level of development, it is even more of one in relation to science—an activity distinguished mainly by the fact of its internationalism. Science cannot exist in isolation and in one sense the endogenous development of science is a self-contradictory concept—though maybe one which is preferable to external technological domination. According to some critics, the 1970s' fight to avoid dominance by the rich and powerful has led to some curious and retrogressive moves in development philosophy.

For example, Dr Geraldo Cavalcanti, Brazil's ambassador to Unesco, has been quick to point out the similarities between 'gap theories' of development, dating from the 1950s and the early 1960s, and ideas about the endogenous development of science and technology. The first failed to work because, when economic forces were left to act freely, the gaps did not reduce but widened. Yet today, though we have a much richer view of economic development, we still deplore the gap between levels of scientific development in developing countries and the rest of the world.

I think we should have a richer concept of the problem of scientific and technological development, stressing the pluralistic and qualitative aspects of progress in these fields . . . Is it not possible to conceive that the diverse technological needs of developing countries, the social contexts in which they appear, the cultures they have to serve, could produce alternative scientific solutions, better adapted to the diversity of national conditions? Would it not be possible to envisage a pluralistic development of science and technology? . . . It seems evident that as long as we keep to a linear concept of scientific and technological progress, we shall sink deeper into frustration and despair. On the other hand, a world which respects and encourages diversity, which seeks not uniformity, but compatibility and integration, will have a greater chance of making Man the harmonious creature he should be.[12]

It could be that this kind of pluralism in science develops more quickly than we anticipate. Ceratinly there is action in the world of Islam to develop its own centres of scientific excellence, ones which will undoubtedly concentrate first and foremost on issues of relevance to the Islamic world. Early in the 1980s the newly created Islamic Foundation for Science, Technology, and Development (IFSTAD), based in Jeddah, Saudi Arabia, was given two priority programmes by the Islamic summit, held in Mecca in January 1981.

The first was to 'develop institutions for studies, research and publications on Islamic ethics and values in science and technology for development' and the second was to reverse the brain drain and selective migration, and find better ways of using indigenous talent.[13] Anyone watching such developments cannot but remember that for many centuries, while Europe lay in the Dark Ages, Islam was the centre of the world of learning. It laid down the basis of a science which everyone today regards as Western—yet which has traditions extending far back in time, even beyond the Islamic world of the ninth, tenth, and eleventh centuries, to the China of 2,000 years ago.

It is odd, now, to have to reflect that throughout the past two millenniums knowledge and learning have always been held in high esteem somewhere in the world—not for their utilitarian value, their relevance to development (which was itself an unknown concept), but for their own sake. For the moment, it would seem, we have become more materialistic, seeing in knowledge little more than a useful tool for the future. Just how useful that view of scholarship will turn out to be remains to be seen—but it is a frighteningly young concept, hardly a few decades old. Learning itself is almost as old as man—and, stranger still, used to be the yardstick by which civilization (or, as we would say today, level of development) was measured. Put like that, it is hard indeed to see, in Casimir's words, how knowledge can be applied to development. Historically, development has always been a measure of knowledge.

Part II · Trends

We must not ask ourselves where our science and technology are taking us but rather how can we manage science and technology so that they will help us get where we want to go.

Manfred Lachs, 1982

Since the 1960s, much attention has been paid to forecasting the future of science and technology. We do so here with some diffidence because the point which Lachs, a past President of the International Court of Justice in the Hague, stressed at the London public hearing on the environment[1] in 1982 is a sound one. It is more important to find ways of persuading science and technology to provide us with what we need than it is to try to guess at what science and technology may come up with if left to their own devices.

As we have seen, these two different ideas stem basically from the 'social pull' and 'science push' models of how science, technology, and society interrelate. But even if the 'social pull' model seems socially and politically preferable, it is still useless to ask for results from science and technology in areas which have not reached fruition. In other words, before specifying our demands, some ideas of what is and what is not possible are in order.

The aim of Part II of this book is to display these possibilities— to offer the reader a broad, sometimes a very broad, canvas of the future. By describing what now appear to be the most active areas in the basic sciences and in technology, and providing an assessment of the state of the environment, we can gain some idea of the boundaries of possible progress during the next decade and a half.

It should be stressed that these are only the boundaries—the area within is not likely to be completely filled in within such a short time period. More likely, advances will come in unrelated patches; while large areas may be painted in, even larger ones are likely to be left blank; and of course as detailed knowledge accrues in any particular area, it shifts outwards the boundaries of possibility, enlarging the canvas yet further.

The boundaries may also be changed by a factor yet to be mentioned in this book—surprise. In the heyday of technological forecasting—the 1960s—there was much talk of creating 'surprise-free futures'. Few people now believe that this is possible; the history of the science/society interaction suggests that surprise is an almost constant feature of change. However, as the United Nations Environment Programme has carefully pointed out,[2] this does not condone a fatalistic attitude. On the contrary, planning for uncertainty—which human beings have always had to do to some extent—is a crucial component of our attitude towards the future. While we cannot foretell the future, nor remove its capacity to produce the unexpected, we can at least prepare and plan for the certainty that the future will never be quite as we envisage it.

5 Basic sciences

Most students of the dynamics of science and technology now refuse to make rigid distinctions between the two. Quite rightly, they see a pair of inseparable twins, each highly dependent on the other. To talk of the future of the basic sciences on their own is therefore somewhat anachronistic—and we do so here only to be able to pin-point those areas of science in which there is currently active interest, and from which the scaffolds of future knowledge may soon be erected.

Biology

Since the 1960s the biological sciences have been undergoing more rapid change than any other field—notably because of the advent at that time of the new science of molecular biology. While interest in that field has continued unabated—and, indeed, in some areas such as the use of recombinant DNA even intensified—new interests and social demands have also concentrated attention on other areas of the life sciences.

Towards the end of the 1960s, concern for the environment gave great encouragement to the study of ecology. One result of this has been a healthy growth in what used to be called whole-animal and whole-plant biology—in contrast to the more recent emphasis on the molecular structure of living systems. This trend has proved fortunate, because as the energy crisis of the 1970s began to bite, it became clear that heavy demands were going to be made on biologists in the general area of biomass.

These two trends have been responsible for the rejuvenation of work in previously rather neglected areas of the life sciences. Partly for this reason, and partly because of the promise of biotechnology, the life sciences are now widely regarded as the leading edge of contemporary science; and it is from them that the results which will most change society—or at least best respond to its needs—are expected during the next decade and a half.

In the past, of course, the life sciences have enjoyed their most major practical successes in the fields of medicine and agriculture; and, indeed, from a purely technical viewpoint the Green Revolution of the 1960s, brought about by the breeding of high-yielding plant varieties, was an outstanding example of practical biology. Nevertheless, the pace of advance in agriculture and medicine was probably slower during the 1970s than in either of the two previous decades. One reason is that those human diseases for which scientific cures are not yet available are proving more intractable than was previously suspected. Almost certainly this is because they arise 'to a considerable extent from the inherent weaknesses, imperfections, and aging of human organisms; and these processes are not so readily changed'.[1]

On the other hand, advances in the field of agricultural research may well be more rapid in the future—partly because social demand is high and partly because the field may be rejuvenated by the advent of recombinant DNA technology. There are two extremely important practical goals. One is the transformation of biological systems to provide increased photo-synthetic efficiency—an advance which could have far-reaching energy implications. The other is to provide all food plants with a mechanism for fixing nitrogen, as legumes do. This would drastically reduce global consumption of expensive fertilizer. While work continues avidly on both fronts, the changes of short-term practical success cannot be described as high.

Human understanding of the molecular mechanisms involved in life forms is far more advanced than was the case even a decade ago, and the pace of advance shows no sign of slackening. These molecular studies have been able to forge satisfactory links with other biological fields operating at higher levels, thus beginning a process of unification which is proving very welcome to life scientists. It is, however, far from complete; in particular, links between molecular biology and the process by which whole organisms develop is far from well established.

While understanding of the way in which a single gene expresses itself, and of what controls the operation of gene expression, is now advanced, the actions of genes in concert is a great deal more complicated. The types of question which need answering include the following. How does a single fertilized egg turn into the 10^{14} cells of the human adult? Why is cell differentiation in plants reversible while it is not in animals? What causes clumps of cells to group together to form the organs of a developing embryo? How does a damaged organ repair itself? And what processes are involved in cell ageing?

Such questions are likely to be fruitful areas for research for many decades. Whether this is true of attempts to link molecular biology with studies of the whole brain and human behaviour is less clear. While many scientists are optimistic, others fear that the formulation of the problem—the linking of molecular genetics to behaviour and individuality—is inappropriate. Indeed, there are grounds for suspecting that in this area a reductionist approach may prove unrewarding.

Molecular biology itself has as yet produced rather few practical applications. This may now begin to change. Use of recombinant DNA technology is likely to lead to the synthesis of a wide range of pharmaceuticals, for instance. And a molecular understanding of such diseases as cancer and the auto-immune disorders may not be far away. Whether genetic engineering will itself prove to be a useful medical technique within the foreseeable future remains more doubtful.

One of the more interesting trends is that the period 1953–80 may have been the heyday of molecular biology. Bernard D. Davis has written:

It seems probable that relatively few major, universal rules of cell organization still await discovery: and with the molecular revolution thus subsiding, molecular biology seems to be losing its identity. In one direction, it is being absorbed into biochemistry and biophysics, as phenomena already known in outline are being examined in ever greater chemical and physical detail. In another direction, it is being integrated into the field of cell biology.[2]

Chemistry

The boundaries of chemistry are less clearly defined than those of either physics or biology—and indeed they overlap with them in many places. Despite this, chemistry occupies a somewhat unique position in the sciences, often providing both the insights and the techniques from which other areas can advance.

The range of interest of contemporary chemistry is very large—extending from the investigation of new chemicals of technological interest on the one hand to the examination of the most detailed facets of molecular architecture on the other. In between lie such current concerns as how catalysts (and their biological counterparts, the enzymes) work, how chemical structure is related to the chemical properties of a molecule, how the synthesis of very complex molecules containing hundreds of atoms is best achieved and how molecules behave under varying physical conditions. In all this the computer is playing an increasingly important role—even down to suggesting the most economic route for the synthesis of individual molecules. Similarly, our understanding of molecular chemistry has recently advanced enormously as a result of increasingly sophisticated methods of analysis such as nuclear magnetic resonance, Auger electron spectroscopy, and many others.

One of the major challenges facing chemists over the next decade and a half is to prepare for important changes in the materials used as the basis of the chemical industry. As the light oils become scarcer, heavy oils, shale-oil, coal, and other materials, including biomass, must provide the basic feedstocks. Coal and biomass have already become important starting materials for the production of synthetic gas and methanol, from which a whole range of other organic compounds can be produced. It now seems likely that over time there will be a continual shift in the nature of the feedstocks used in chemistry—from petroleum-based products now, to coal and gas in the near future, to biomass sources and waste products in the long run. Each shift will make heavy demands on both chemistry and chemical engineering.

It is the efficiency of these processes which will come under closer scrutiny in the next few years, as more effective and more specific catalysts are devised to cope with each step in the reaction chain. The production of liquid fuels from non-petroleum sources is likely to remain an important chemical problem for some time to come.

The investigation of such processes depends vitally on an understanding both of how chemicals react together and of the role played by catalysts in speeding such a reaction. Two important new tools have recently been added to the chemists' armoury which are particularly valuable for studying the kinetics of reactions in which chemical bonds between atoms may be broken and remade in less than 10^{-9} seconds. These are known as molecular beam and laser spectroscopy, and they have been used to study the reactions of two or more molecules together almost on an individual basis. In spite of all their knowledge of the details of chemical reactions in bulk, chemists still know little about what happens at the molecular level, even in the simplest chemical reactions of all.

Similarly, although an individual molecule can be described in terms of quantum mechanics with great accuracy, it remains difficult to predict the properties of any chemical containing more than a dozen atoms. Only for molecules with four or less atoms can all the properties of the molecule be deduced in great detail from theory. However, the reintegration of practice and theory in chemistry—which has been greatly aided by the computer, and particularly by computer models of molecules—is likely to advance rapidly in the near future. This may be significant because, as just mentioned, the energy crisis and petroleum shortages will soon force major changes in the feedstocks used in the chemical industry.

There are some areas of chemistry where the rate of advance has slowed down—for instance, in the production of new agricultural and pharmaceutical chemicals. In fact, the amount of research and the number of new compounds isolated and investigated does not appear to have been reduced (some 6,000 new compounds are still reported every week). However, the rate of introduction of new commercial chemicals is slowing down, and for two reasons: first, it is becoming increasingly difficult to discover compounds which produce novel or improved effects when so many are already known; and, second, environmental and safety requirements for new chemicals are far more stringent now than they used to be, making the introduction of new products more difficult and more expensive.

At the same time, the chemist's ability to synthesize new molecules has improved remarkably, mainly as the result of the discovery of new organo-metallic catalysts which can bring about extremely specific reactions. Using a computer, the synthetic chemist can now draw out the shape of the molecule he needs to synthesize, and ask the computer to suggest, on the basis of previously stored data, step-by-step synthetic routes starting from a suitable feedstock.

This ability, however, still falls far short of that of most living systems which are able to synthesize highly complex molecules with great accuracy and speed. Exactly how this is achieved is becoming one of the major areas of study in macromolecular chemistry. This has led chemists to spend more and more time studying the spatial conformations of macromolecules, because it is becoming clear that the way in which a long molecule arranges itself in space has much to do with the way it reacts with other molecules. Increased understanding of the chemical mechanisms involved in biological processes should eventually lead to major advances in medical and agricultural chemistry as chemists learn to design molecules which will perform highly specific tasks within various living systems.

Of course, many isolated individual problems remain. Two of these, for example, are to find organic semiconductors and to discover materials which will exhibit superconductivity at relatively high temperatures. In a sense these two problems, and a number of related issues, have been on the agenda for a long time now. However, work is steady, if not spectacular.

Physics

While progress in physics has perhaps been less dramatic in the past two decades than previously, there are signs that major new advances—perhaps surpassing anything previously made in the field—are now not far away. These signs come from two different areas: there is a suspicion that the fundamental building blocks of matter may now finally have been identified; and mounting evidence that the theoreticians may soon produce the long-awaited unifying field theories which will unite all the basic forces known in nature.

The fundamental building blocks of matter are now thought to be the two kinds of particles known as leptons (of which there are six) and quarks. Quarks are found in the atomic nucleus and are the subject of intensive research. Before 1974, three kinds of quark were known, each characterized by a quantum attribute, known respectively as up, down, and strange. In 1974 the discovery of the omega minus particle—comprised solely of three strange quarks and with properties corresponding exactly to those predicted in theory—provided vivid practical demonstration of the theory.

These three properties of quarks are sufficient to explain the existence of the group of particles known as baryons. The mesons, however, can be explained only if it is assumed that each meson is comprised of a quark–antiquark pair. Advances in theory in the early 1970s showed that there must, in fact, be a fourth kind of quark, with a property called charm. Once again, the theory predicted the existence of a particle composed of a combination of a charmed quark and a charmed antiquark. This particle, the J meson, was discovered in experiments in 1975. In the late 1970s, the discovery of yet another set of particles proved the existence of a fifth quark, with a characteristic known as bottom.

It now seems highly likely that there must also be a sixth quark, with an attribute known as top. Experiments to find it will soon be in progress, and many physicists believe that this will mark the end of the search for the fundamental building blocks of matter. As much is indicated by the nature of the mathematical group theory which underlies contemporary particle physics. However, should more than six quarks be discovered it may be that, as has already happened a number of times in physics, physicists will have to seek even more fundamental building blocks; the quarks would then turn out to be composed of other, more fundamental entities. Some theoreticians have already begun to try to show that all nature can be explained in terms of just two other particles.

For many years physicists have wanted to unify their subject. Essentially, this means producing a set of theories which will encompass the four fundamental forces known in nature—electromagnetism, the weak and the strong interactions, and the force of gravity. In 1974 a major success was achieved in that theories unifying the first three of these were announced. In particular the electromagnetic and weak interactions were unified in a theory which predicted that certain fundamental particles, two charged W particles and a neutral Z particle, must exist. These particles would play a similar role to the one played by the photon in the quantum electrodynamic theory of electromagnetism. The theory predicted the properties of these new particles very precisely. Early in 1983, all three particles were found in experiments at CERN in Geneva (an international particle-physics laboratory which Unesco helped to found in the 1950s). When the force of gravity can also be brought into this scheme, and if the quarks do indeed turn out to be limited to six, physics will itself have taken a quantum jump ahead.

Work on the behaviour of matter at various different levels continues to form the corner-stone of physics. The nucleus continues to be studied with avid interest and the behaviour of plasmas—hot, ionized gases—has been the focus of greatly increased work as a result of the energy crisis of the 1970s. The goal here is the attainment of controlled thermonuclear fusion in the laboratory, a goal which when achieved will provide the possibility of producing unlimited energy from hydrogen isotopes found in sea-water. So far success has proved elusive but physicists expect it within a decade or so. The problems of scaling up the process to commercial level, however, are enormous—far more complicated than were those surrounding nuclear fission. For this reason most experts do not expect to see commercial fusion reactors operating on the planet before the year 2020, at the earliest.

The study of more-condensed forms of matter continues to excite physicists, particularly in two areas: the study of surfaces (for instance, the interaction of gases with surfaces); and the study of amorphous solids. These solids, which exhibit far-less-ordered structures than other materials, are being intensively investigated for they offer the possibility of being used in place of more-ordered materials in electronic devices, with ensuing advantages of cheapness and much decreased sensitivity to contaminants—one of the things which makes the manufacture of electronic components from, for instance, silicon crystals relatively expensive.

Two factors have combined to shift the focus of much solid-state research. First, a growing shortage of such materials as chromium and cobalt has produced great interest in making physical analogues of these materials using the technique of ion-beam implantation. Second, dramatic advances have been made in the technologist's ability to fit increasing numbers of electronic devices on to a single silicon chip. Between 1970 and 1980 this rose by a factor of 1,000 (progress in using solid-state devices in information technologies is described in more detail in the next chapter). These two examples provide dramatic cases of the 'society pull' model of scientific advance. Writing about the current state of the art of contemporary physics, Professor D. Allan Bromley of Yale University, comments:

The ultimate consequences of such developments are beyond our present imagination, but already they have entirely changed the character of much of physics. We can now ask questions of a complexity and sophistication that would have been out of the question even five years ago.[3]

No similar incentives have been operating in the field of fluid physics—although the relatively recent acceptance of the idea of plate tectonics, in which large plates of the Earth's crust slide over one another, thus explaining many of the previously unsolved problems of geophysics, has certainly stimulated new work in fluid mechanics. The phenomenon of turbulence is once again being intensively studied, even though previous attempts to do so were only partially successful. Encouraging progress has been made both here and in the area of 'quantum phenomena' such as superfluidity.

Many of the more bizarre discoveries of physics—the strange behaviour of matter under extreme conditions—are being made by mostly theoretical studies in astrophysics. Quasars and pulsars, black holes, black-body radiation, and neutron stars all lead the physicist's mind to dwell on aspects of physics at the extremes of the human imagination. For example, if primordial black holes do exist in the universe as remnants of the Big Bang, then they would have a mass equivalent to that of Mount Everest packed into a space no larger than that occupied by a hydrogen atom.

In astronomy itself, the rate of advance over the past twenty years has been outstanding. Instead of simply viewing the universe by means of visible radiation, the wavelengths available to the astronomer now extend from 10^{-14} to 100 metres. And in addition to the traditionally studied stars and galaxies, there is now a whole range of new objects to consider. As Vera C. Rubin has put it, 'An exotic menu of astronomical sources is now routinely available for study, and the universe is known to be immeasurably richer, more varied and more violent than would have been dreamed even 20 years ago.'[4]

Among the major events of the next decade or so will be the launching of the Cosmic Background Explorer (COBE) which will measure the primordial background radiation in the millimetre and submillimetre range from space. Such measurements may reveal whether the background radiation really is 'fossil' radiation from the Big Bang which marked the beginning of the universe, as most astronomers believe. They should also reveal something of the nature of the very early universe, such as its speed of rotation, and act as a speedometer for measuring the motion of our own galaxy.

During the 1980s the Space Telescope will be launched, providing astronomers with a high resolution optical telescope outside the Earth's atmosphere. Among the many enormous advantages this is likely to bring is a greatly improved measurement of the rate of expansion of the universe. The advent of these two major astronomical instruments echoes what has been going on in many other fields of physics where instrumentation is playing an increasingly important role: often, the frontiers of today's physics are determined by the presence or absence of appropriate instrumentation. And there is constant, two-way communication between instrumentation and these frontiers; just as new instruments make it possible to extend the frontiers, so too do advances in physics extend the frontiers of the instrumentation which is available, to the mutual benefit of both.

Attempts to chart the distribution of matter in the universe have shown that most of the matter in it is, for one reason or another, invisible. Its presence is now being revealed by its gravitational effects on other objects which can be seen. One estimate is that as much as 90 per cent of the universe is currently unseen. The distribution of matter in the universe is also being more closely studied now that astronomers are beginning to concentrate on the clusters which galaxies form, rather than on the galaxies themselves or the stars they contain. The 1980s are likely to be an epoch of 'cluster study'.

Whereas this may seem to be a rarefied world, the advent of space technology—and now the Space Shuttle itself—is likely to provide important new insights into the workings of the universe. The shuttle will open up new regions of the spectrum with which to view the universe, making new kinds of observation possible and undoubtedly changing our conceptions of the universe a great deal. It is one of the most curious phenomena of modern physics that its leading edges, dealing with the macro-world of the universe and the micro-world of subatomic particles, are both explained in terms of the same ideas and concepts; indeed these ideas and concepts frequently depend critically upon one another. Physicists now believe that they can trace the evolution of the universe from a date of 10^{-43} seconds up to the present in some detail. Steady-state theories of the evolution of the universe, which gained some currency during the 1960s and 1970s, have now been largely replaced by theories which invoke an expanding or inflationary universe.

6 Natural systems

A few thin metres of soil, a few miles up into the sky and a similar depth down into the oceans encompasses virtually the whole of the biosphere in which we and other living things can survive . . . For an increasing number of environmental issues, the difficulty is not to identify the remedy, because the remedy is now well understood. The problems are rooted in the society and the economy—and in the end in the political structure.

Barbara Ward, 1914–81

The concept of the Earth as a planet, while old to astronomers, is recent to the general public. It dates first from the advent of the space age and second from the rise in environmental awareness which began in the late 1960s. These have had a profound influence on the ways in which the planet is now perceived. Where once we had geology, geophysics, and geography, we now have a more holistic concept which links these fields together: that of the Earth's natural systems—the crust, land, water, oceans, and atmosphere—which together comprise what we know of the planet. Though we may study these systems as separate entities, we no longer regard them as distinct. They form a continuum, and scientists are increasingly studying them as such.

One of the reasons is that as human intervention in the great natural processes has become more significant, an urgent need has arisen to understand these processes in detail. For example, the prospect of greenhouse heating of the planet due to the accumulation of carbon dioxide in the atmosphere has led to many new studies of the carbon cycle—of the ways in which carbon is cycled through the natural systems. What happens to carbon when coal and oil are burnt? How much stays in the atmosphere and for how long? How much is dissolved in the oceans? How much returns to land as carbonic acid? How large are the environmental reservoirs of carbon, and how fast is carbon transferred from one to another?

Many questions of this type were effectively answered during the 1970s, though there are still gaps in our knowledge. Thanks to the initiative of SCOPE, the Scientific Committee on Problems of the Environment (a component of the International Council of Scientific Unions), special units were set up during the 1970s to study several of the major biogeochemical cycles:[1] one on nitrogen in Stockholm; two on carbon, the main one being in Hamburg, with a sub-unit in Stockholm; and one on sulphur in Pushchino in the Soviet Union. During the 1980s further units may be set up to study the biogeochemical cycling of metals, increasing emissions of which are presenting toxic hazards.

These and other studies are concerned basically with the question of how the natural systems work and of what changes are occurring in them as a result of human action. In this sense the science of natural systems falls midway between the basic sciences and the technologies: basic studies reveal the inner workings of the systems and this knowledge enables us to improve our management of them. In other words, the sciences involved are neither academic nor applied, but they certainly have direction. This concept has proved extremely difficult for those who still insist on the linear concept of research, development, and application. It is, however, part and parcel of the process described elsewhere in this book in which distinctions between understanding and application are breaking down.

The Earth's crust

The past two decades have been an exciting time for those bent on understanding the processes at work in the Earth's crust. A single theory—that of plate tectonics—has helped to unify many different disciplines in the Earth sciences, providing an overall model capable of explaining most of what happens in the crust.

In fact, our ideas about the nature of the Earth have changed almost beyond recognition since 1960. At that time even the idea of continental drift was widely disputed and it was assumed that the continental and oceanic crusts were of equal age. Today, few scientists indeed still quarrel with the idea of continental drift; and though the continental crust is thought to be around 4,000 million years old, the oceanic crust is known to be much younger —indeed, neither the Atlantic nor the Indian Ocean existed 200 million years ago.

The concept of an old and stable oceanic crust has been replaced by one in which we now know that all the oceanic crust has been formed or renewed in the past 200 million years. As one commentator has put it, 'It is such wholesale renewal of the 70 per cent of the Earth's surface that is occupied by oceanic crust in less than one-twentieth of the geological age of the Earth that is the most fundamental change in our view of the Earth.'[2]

This sudden swing in geophysics was due to several, almost coincidental factors. During the late 1950s, palaeomagnetic studies of rocks indicated what were called 'polar wandering paths'—it appeared that the location of the magnetic poles had wandered about over time. Indeed, they had—but the studies also showed that the way the poles wandered over time appeared different from each continent. It was this piece of evidence which finally convinced most geophysicists of the fact of continental drift.

The result was that the earth sciences were, by 1960, poised for a revolution in geological thought. The proof of continental drift was to lead to a new concept of global dynamics and plate tectonics—a broad, synthetic approach that linked sea-floor spreading, volcanism, earthquake activity, mineralization and mantle evolution into one all-embracing theory. The 1960s closed with Man's landing on the Moon and this also enabled us to see the Earth as never before—in perspective and from a distance—sharpening the realization that this is an active planet far different from its satellite.[3]

Subsequently, the Deep Sea Drilling Project of 1965–75, followed by the International Phase of Ocean Drilling, provided material from several hundred deep boreholes in the oceanic crust which revealed the youth of the crust. These results confirmed the theories of sea-floor spreading which were first suggested in the 1960s. The proponents of continental drift then set to work to establish whether or not there really was an exact fit between the coastlines of Africa and South America—earlier studies had been shown to involve discrepancies of at least 1,000 kilometres. Using a computer to do the geometry, and fitting not the coasts but the edges of the continental shelves, this discrepancy was reduced from 1,000 km to only about 50 km, and one of the key objections to continental drift was finally removed. At the same time the theory of plate tectonics was being evolved—a theory which would provide a mechanism with which to explain not only continental drift but almost every other major geophysical event occurring in the Earth's crust as well.

This theory assumes basically that the crust is composed of a number of large plates, moving slowly in relation to one another by a few centimetres a year. This concept clearly explains the facts of continental drift, the occurrence of volcanic activity, and the reasons for earthquakes. The latter, for example, result from friction between two overlapping plates. Stresses build up as plates interact with one another, until they reach the point at which something has to give. The sudden release of pressure produces shock waves in the surrounding area and the plates move in a matter of minutes distances of as much as 10 metres—a movement which would normally take decades or centuries.

Plate tectonics has also produced a general understanding of the processes which give rise to mineral deposition. The theory is helping scientists understand how climate has evolved with time, solving long-standing puzzles in the field of vertebrate evolution and providing oceanographers with new information about the ways in which ocean sediments are formed.

All this is of great potential economic value. At the beginning of the 1980s, however, the major problem was that plate tectonics was still essentially a macro-theory. While it could provide information on the type of conditions needed, say, to produce an earthquake, stimulate volcanic eruptions, or form high concentrations of useful minerals, it could do so only in general terms. The theory could be used to predict where such things would not happen or would not be found; and it could be used to predict where they could happen or might be found. But it could not predict whether they would. The sophistication of the theory to the point at which it can explain and predict at the micro-level and thus be of use to economic geologists and others is one of the major goals of the future.

One of the most important needs in the field is for accurate means of earthquake prediction. In fact at least three earthquakes have now been accurately predicted, the most successful prediction being that of the Haicheng earthquake in February 1975. Thanks to a major Chinese programme begun in 1968, tens of thousands of volunteers and scientists had been measuring such things as water-levels in wells, animal behaviour, and changes in ground electrical currents on a systematic basis. These and other observations led to the public prediction of the Haicheng earthquake five hours before it occurred. The population was successfully evacuated in time, and tens of thousands of lives were saved.

During the next decade our ability to predict earthquakes is likely to improve considerably. The process, however, may be expensive in terms of the great number and frequency of the observations which have to be made. This means, in effect, that it may prove economic only in areas of very high earthquake risk. However, this is an area of vital importance to humanity because the loss of life and property from a large earthquake—which can physically move portions of the land surface by several metres—can be enormous. Some 70,000 people were killed in the Peruvian earthquake of 1970. Earthquakes remain the most damaging of all natural disasters.

The land

During the past decade a series of studies of land-use management has led to a greatly improved understanding of the effects of human activity on the 150 million square kilometres of the Earth's land surface. There have been two main motivations for this work:

- the need to increase food production, both to meet current needs and to prepare for future increases in world population
- increasing concern with land deterioration as a result of environmental pressures. Over the past decade desertification and deforestation, for example, have been continuing apace.

Only 11 per cent of the Earth's land surface is suitable for cultivation. Most of it is already being used. Currently, each hectare of good arable land supports 2.6 people. Over the next two decades this figure will have to be substantially increased. The possibilities of finding more arable land are limited by the conditions which exist over the rest of the Earth's surface, of which:

- 6 per cent is permanently frozen
- 10 per cent is waterlogged
- 22 per cent has little or no topsoil
- 23 per cent suffers from salinization
- 28 per cent is too arid.[4]

One of the major problems is that as agriculture is extended to increasingly marginal land, and good land is used more intensively, food supplies become more prone to disruption. Droughts, storms, and other normal weather variations take a greater toll.

Arable land is currently under threat from many different sources. Every year between 5 and 7 million hectares of good land are built on. Between 1960 and 1970 Japan lost no less than 7.3 per cent of its agricultural land to buildings and roads.[5]

Soil erosion is also on the increase. In India, more than half of all agricultural land is subject to degradation. Some 6,000 million tonnes of soil are lost every year from just 800,000 square kilometres of land.[6] They carry with them more than 6 million tonnes of nutrients, more than the amount that is applied in the form of fertilizer.[7]

The use of intensive methods of farming is also leading to loss of organic matter in the topsoil and the buildup of both toxic chemicals and salts in the soil. At the current rate of land degradation, it is estimated, as much as one-third of all good agricultural land could be lost within the next two decades.[8] The problem is aggravated by the fact that so much natural organic fertilizer, instead of being returned to the soil where it is badly needed, has to be burnt for fuel. Currently, an estimated 400 million tonnes of dung and crop residues are used as fuel, almost entirely in the developing countries.[9]

The UN Food and Agriculture Organization claims that food production needs to be increased by 60 per cent over the next twenty to thirty years. This could be achieved by intensifying production and bringing another 200 million hectares of arable land—an increase of 1.3 per cent—into production by the end of the century. But according to FAO this would only just compensate for the amount of land lost by degradation if current rates continue unchecked.[10]

Desertification is one of the major problems. Aridity affects nearly one-third of the land surface. It is currently threatening the lives of the 80 million people who survive on the 19 per cent of the land under attack from desertification. Some 20 million hectares now deteriorate annually to the point where they stop yielding an economic agricultural return. The cost of lost production has been estimated at $26,000m. a year.

Desertification is not being produced by climatic change but by overcropping, overgrazing, and salinization. It can also be produced by deforestation, which exposes soil to wind and rain, resulting in sudden soil erosion and flooding. But as the soil disappears so does the ability of the land to trap and retain moisture. The desert begins to get the upper hand.

In the late 1970s work by FAO and the United Nations Environment Programme revealed that rates of deforestation were not quite as bad as had been feared. Nevertheless, it is thought that for the period up to 1985 the open forests will disappear at a rate of 3.8 million hectares a year and the closed forests at 7.5 million hectares a year. On present trends 9 countries will have destroyed all their closed forest within 30 years, and a further 13 nations within 55 years.[11]

The implications are serious. For one thing, deforestation inevitably adds to the carbon dioxide burden in the atmosphere, increasing the chance of climatic change as a result of the greenhouse effect. For another, exposing thin tropical soils to the elements often leads to their eventual disappearance. The land which supports tropical forests is often unsuitable for agriculture, even though this is still the commonest reason for felling in the tropics. Thirdly, the increasing scarcity of timber in tropical regions is leading to fuel-wood problems for hundreds of millions of people. And, finally, the forests contain most of the world's genetic resources, which are now under increasing threat.

Exactly how many species the Earth supports is still unknown —estimates vary from 3 to 10 million. However, only 1.5 million have been scientifically recorded. The rate of discovery is still swift: in the African rain forests alone more than 200 new plant species are being recorded each year. What is clear, however, is that many species are disappearing. Current estimates are that 2,000 vertebrates are in danger of extinction as are 10 per cent of all the flowering plants known to science—about 25,000 species. Of the 145 indigenous cattle breeds found in Europe and the Mediterranean, no less than 115 are now threatened with extinction.[12]

It is vital that as many species are possible be saved. One reason is economic. We depend for new varieties of crops and animals on the wild gene stock available from which to cultivate improved strains. As an example, in the 1860s an insect feeding on the roots of vines destroyed nearly every vineyard in Europe. Only the discovery of vines in the United States which were resistant to the insect made it possible for the vineyards of Europe to be restocked, by grafting European vines on to American roots—a practice which continues to this day.[13]

In addition, almost all our agriculture and much of our medicine depend on strains which have been successfully adapted from the wild gene pool. In the United States alone, the value of medicines sold each year which are extracted from various plants is in the region of $3,000m. And as UNEP's Executive Director, Dr M. Tolba, said in November 1981, 'In Europe there is now no such thing as a home-grown meal. Agriculture (and pharmaceutical and certain other industries) depend on infusions of fresh germ plasm imported from the genetic powerhouses which, except for a part of the Mediterranean, are all located in developing countries.'[14]

The importance of this is not simply that the wild gene stock provides something to fall back on in times of emergency. It is, in fact, being constantly used to improve species which are increasingly vulnerable to attack from different pests which are becoming more and more resistant to pesticides. Indeed, according to the World Conservation Strategy, the average lifetime of wheat and other cereal strains used in Europe and North America is now only 5–15 years.[15] The issue, in other words, is of much more than academic or emergency importance.

The discovery of new species also holds out the promise of new products. Rubber and quinine both came from the discovery of new trees. It is quite possible that new products of equal importance may yet emerge from the gene stock which is still waiting to be discovered, particularly in the tropical rain forests. The armadillo, for example, has recently been discovered to be the only animal other than man capable of contracting leprosy. As a result of this, a leprosy vaccine is now being developed. And in view of the growing importance of biotechnology, the preservation of microbial strains of life is also becoming urgent. This is a field in which there has been substantial international action. Between 1975 and 1980 five microbial resource centres were established in developing countries, and the World Data Centre on Micro-Organisms can now provide information on 24,700 microbial strains in 64 countries.

But as well as economic reasons, there are moral ones which behove us to preserve as much of the living heritage as possible. This case has been clearly made by the World Conservation Strategy:[16]

Human beings have become a major evolutionary force. While lacking the knowledge to control the biosphere, we have the power to change it radically. We are morally obliged—to our descendants and to other creatures—to act prudently. Since our capacity to alter the course of evolution does not make us any the less subject to it, wisdom also dictates that we be prudent. We cannot predict what species may become useful to us. Indeed we may learn that many species that seem dispensable are capable of providing important products, such as pharmaceuticals, or are vital parts of life-support systems on which we depend. For reasons of ethics and self-interest, therefore, we should not knowingly cause the extinction of species.

Much work remains to be done. More determined efforts to classify the remaining unknown species are needed, improved legislation is required to protect and preserve endangered species—a realistic possibility as the successful fight to save the whale has shown—and many more national, regional, and international gene banks are needed as information centres and as storehouses of valuable genetic material. All this requires powerful international action by the scientific community.

Since the publication of the World Conservation Strategy in 1981, a global plan has existed for creating and maintaining a healthy biosphere. Implementing that plan, however, is another matter. The various United Nations agencies which share a responsibility for the environment have already made good progress in organizing a concerted attack on the problem. Their efforts are largely inspired by the Unesco Man and the Biosphere (MAB) programme which was begun in 1971.[17]

At that time, a great deal was known about the environment but this knowledge was rarely applied in the practical, day-to-day management of the Earth's resources. For example, the potential problems produced by large dam projects were well known: all too often since the 1950s such projects had led to siltation, soil erosion, salinization, excessive water evaporation, disruption of fishing, displacement of local populations, the spread of schistosomiasis, and other problems.

These problems were extensively documented.[18] Yet decision-makers still continued to build large dams, apparently oblivious of the practical drawbacks which in the long term could well offset any short-term advantage. Unesco reasoned that if this were the case there must be obstacles which prevented the existing knowledge from being used, specifically:

- the knowledge might exist, but not in a form suitable for decision-makers and operators;
- the knowledge might be too sectoral in nature to be useful in solving everyday, practical problems;
- the knowledge of specific issues, resources, and ecosystems might be insufficient to provide a sound basis for action.

MAB was therefore conceived as a programme designed to overcome these deficiencies in environmental knowledge. Apart from increasing knowledge of environmental management techniques very considerably, MAB has left a legacy in the form of its 'biosphere reserves' which could prove of great importance in helping to implement the World Conservation Strategy.

Biosphere reserves are designed to fulfil a number of functions.[19] They are intended to protect the flora and fauna, and the genetic diversity, of an ecological area of special interest as well as to provide a place for research, education, and training in the relevant sciences. They are linked together in a world-wide network and the computerized MAB Information System provides information on both the nature and the activities of each biosphere reserve.

Ideally, each biosphere reserve consists of a core area, surrounded by an inner buffer zone and an outer buffer zone. Typically, the core area remains undisturbed and uninhabited. Activity in this area is limited to conservation and there is a minimum of human interference. The core area serves as a baseline against which to monitor environmental changes elsewhere.

The first buffer zone contains the research station, plus other human settlements. It is used for education and training, and for 'manipulative' research on conservation and ecosystem management. There may be traditional land-use activities— logging, grazing, and fishing, for example. The second buffer zone carries this trend a stage further. In it, experiments may be conducted on alternative systems of land use; it is likely to be actively managed for the benefit of local populations, and it may even have to be 'reclaimed' for them.

When the system of biosphere reserves was set up, Unesco made an attempt to define what is meant by a 'representative' ecosystem. A global classification system was set up based on the concepts of 'provinces', 'biomes', and 'realms'—each category representing an ecosystem of increasing generality. Altogether, 193 biogeographical provinces were identified, belonging to 14 types of biome in 8 biogeographical 'realms'.[20]

By 1982 there were 214 biosphere reserves in 58 countries. However, there was a great deal of duplication of types of environment covered—only 91 of the 193 provinces had been covered, the major gaps being in tropical moist forests, warm arid zones, coastal areas, and traditional man-modified landscapes.

However, the potential role of biosphere reserves in world

conservation is very large. In contrast to the idea of national parks, biosphere reserves can stimulate the active involvement of local populations in protecting and developing their own environments. This is an important and so far neglected area in conservation. In the past, the establishment of conservation areas has often alienated local populations, whose opinions were not solicited and whose life-styles suffered severe disruption as a result. The existence of many of these conservation areas is now threatened by an expanding and hostile local population. As Michel Batisse writes:

The biosphere reserve constitutes a technique, among others, to reverse this very dangerous trend. Experience already shows that when the populations are fully informed of the objectives of the biosphere reserve, and understand that it is in their own and their children's interest to care for its functioning, the problem of protection is largely solved. In this manner, the biosphere reserve becomes fully integrated—not only into the surrounding land-use system, but also into its social, economic and cultural reality.[21]

Water

The planet's water resources[22] amount to some 1,500 million cubic kilometres, or 10^{18} tonnes, nearly all of it—actually about 97 per cent—stored in the oceans. Of the remaining 3 per cent which is found on the continents and in the atmosphere, most is in the form of ground water (22 per cent) or ice (77 per cent). This leaves only about 516,000 km^3 of water to participate in the hydrological cycle of evaporation and condensation. It is this portion which is 'renewable'.

The main details of the enormous engine of the earth's water cycle are now well known. Each year solar energy evaporates 445,000 km^3 from the oceans and 71,000 km^3 from the continents. This circulates in the atmosphere and is then precipitated over the sea and the land but in a different proportion to the evaporation: 412,000 km^3 falls on the sea and 104,000 km^3 on the land. This leaves the continents with an annual surplus of some 33,000 km^3—the potential source of hydropower, incidentally—which runs off the land and into the oceans.

This surplus is not, of course, evenly distributed between the continents or even within them. The humid tropical regions receive the most benefit, while in some years the most arid regions may receive no rainfall at all. This disparity in water supply correlates closely with levels of socio-economic development, for adequate fresh water is critical to human survival. Although most countries do, in fact, receive more than sufficient rainfall for the basic needs of their populations, catching it and storing it can be very expensive in regions where precipitation is low or very variable. The uneven distribution of rainfall is dramatically illustrated by the fact that fifteen of the world's largest rivers carry a third of the global run-off, and one—the Amazon—carries 15 per cent.

The human population has to fit itself in with this natural cycle in order to obtain its fresh water. While this has been done without enormous problems in most areas of the developed world, the developing countries, with large populations and generally lower rainfall, are in serious trouble. A survey by the World Health Organization in 1975 showed that 1,200 million people—62 per cent of the population of developing countries, excluding China—were without adequate fresh water. The situation was worse in the countryside than in the city: four-fifths of the rural population and one-quarter of the urban population have inadequate supplies.

This means that in many countries the women and children spend most of their day fetching water. However, the implications of such shortages extend far beyond mere inconvenience. Four-fifths of the diseases of the developing countries are linked either to dirty water or lack of sanitation. Diarrhoea, caused mainly by bacteria transmitted via dirty water, still kills something like 20,000 children every day.

Progress since 1975 has been slow. While the proportion of people without adequate fresh water has decreased, the absolute numbers have increased. By 1980 a quarter of the urban population of developing countries still had problems, compared to 71 per cent of the rural population. But 100 million more people had to drink dirty water in 1980 than in 1975, and 400 million more than in 1975 had no sanitation.

The decade from 1981 to 1990 has been officially established as the International Drinking-Water Supply and Sanitation Decade. The goal is to supply everyone with adequate drinking-water and sanitation by the end of the decade, but there is no hope of the target now being met. To do so would mean providing half a million people a day with new or better services and increasing spending by at least five times; the overall cost would be $80m. a day for a decade, according to one estimate.

Drinking-water and sanitation, though, are only two of the uses made of fresh water. Irrigation water is, in many areas, in equally short supply and is often the bottle-neck to increasing food production. The quantities needed are vast and more research on the specific needs of crops for water, and on means of supplying it with minimum waste, is badly needed. Industry in general and power-stations in particular also consume large volumes of water; in some developing countries lack of water for industrial use is now actually preventing development objectives from being attained. In some developed countries cooling-water for new power-stations is becoming extremely hard to find.

The major freshwater problem is one which does not require further research—as yet. It is mainly a question of the money and politics needed to link the people with the supplies. But additional problems which will require scientific attention are certainly on the horizon. One is the increasing pollution of freshwater sources. Although the hydrological cycle is a natural purifier of the water system, and human techniques for purifying water are cheap and efficient, neither can necessarily deal with huge increases in the rate of pollution. If these are sufficiently high, it is almost inevitable that one man must drink another's dirty water before there is time for either natural or man-made systems to purify it.

The freshwater system, like other natural resources, is also under pressure from high demand. Ground water is being increasingly used both for irrigation and drinking because it is approximately 3,000 times more abundant than the surface water which can be found on the Earth and because it is usually a great deal cleaner. However, this part of the water resource is not strictly renewable. Like the fossil fuels it is being renewed, but only at rates which are orders of magnitude smaller than the rate of consumption. Hence ground water is being mined and in some cases cannot be replaced. As underground aquifers dry up, they may cause ground subsidence.

Pollution is also beginning to reach even these fairly remote water sources. The main problem is agricultural chemicals, particularly nitrate, but pollution from man-made wastes is also causing problems. Underground aquifers, of course, are not even potentially self-cleansing, as are rivers and streams.

Several other hydrological problems await further research during the 1980s. More detailed studies are required of the effects of human activity on the hydrological cycle, the influence of climate on water resources, and means of assessing water resources in arid and semi-arid zones.

Fortunately, research in hydrology is already internationally co-ordinated, thanks to a major programme which dates back to Unesco's first entry into the field in 1950, when the organization launched a research programme on the arid zones. This was followed in 1964 by the International Hydrological Decade, during which major advances in our understanding of the hydrological cycle were made.

The practical issues of water management, however, were still insufficiently studied. On 1 January 1975, therefore, Unesco launched the International Hydrological Programme (IHP), designed specifically to improve water resource management and planning, the assessment of water resources, and the evaluation of the impact of human activity on the hydrological cycle. Education and training were an important part of IHP. Over the next five years, 1,500 specialists were trained in IHP courses, 54 hydrology projects were successfully executed, and 35 scientific reports on hydrology were published.

A second phase—IHP–II—was carried out during 1981–3. The third phase of the programme, IHP–III, is now in operation and differs markedly from the preceding phases. IHP–III[23] is basically much more geared to concrete applications and to involving both decision-makers and the general public in those hydrological matters which affect them. Greater use is to be made of regional seminars and workshops and of pilot and demonstration projects. The programme will cover the period 1984–90 and is designed specifically to fulfil the recommendations of the United Nations Water Conference (1977) Action Plan; more generally, it is hoped that IHP–III will help attain the goals set forth in Vienna by the United Nations Conference on the Application of Science and Technology for Development (1979).

The oceans

Ocean resources have become one of the major issues of international law and the potential of the ocean as a source of food, energy, and minerals is becoming increasingly important. While coastal pollution is serious, particularly near the mouths of large rivers in industrial countries, the mid-ocean remains largely free from the effects of human action.

Most scientific studies of the ocean system relate ultimately to one of three factors: pollution, production of food, energy, and minerals, and the role of the oceans in global phenomena (particularly weather and climate). That oceanographic research is today rather well co-ordinated internationally is due mainly to Unesco's initiative in setting up the Intergovernmental Oceanographic Commission (IOC) in 1960.[24] Its job is to define the problems which require international co-operation in oceanography if they are to be solved, and then to develop programmes to solve them. With more than 100 member-nations, the IOC is now the principal focus of international marine research.

Although the International Indian Ocean Expedition (1959–65) was in existence before the IOC, the latter soon came to play a major role in co-ordinating research on the Indian ocean. The results were published in five major atlases, and a great deal of new information was discovered about the relationships between monsoons and surface currents. Surveys showed that less than 0.1 per cent of the primary production of the Indian Ocean was used by man, and that a tenfold increase in catch could be expected with conventional technology.[25]

During the 1960s, the IOC co-ordinated major research

programmes in the tropical Atlantic, the Caribbean, and on the Kuroshio current flowing along the east coast of Asia. It also set up a programme for research in the Mediterranean, and encouraged more research in the Antarctic and along the north-west coast of Africa.

By 1970, the IOC had published its now famous Long-term and Expanded Programme of Oceanographic Exploration and Research (LEPOR). LEPOR was aimed not only at increasing understanding of what is going on in the oceans but also at 'the goal of enhanced utilization of the ocean and its resources for the benefit of mankind'. Basically, LEPOR recommended some fifty research programmes in six main areas:

- problems of ocean/atmosphere interaction, ocean circulation, and tidal waves;
- living resources and their relationship to the marine environment;
- marine pollution;
- geology, geophysics, and marine resources below the ocean floor;
- the Integrated Global Ocean Station System (IGOSS);
- international investigations in specific regions.

LEPOR is an ongoing programme designed to be carried out over several decades. In order to get it off to a good start, the IOC established the International Decade of Ocean Exploration (IDOE) to last from 1971 to 1980.

As a result of the UN Conference on the Human Environment (Stockholm 1972), and at the request of the UN Assembly, IOC initiated an important programme called Global Investigation of the Pollution of the Marine Environment (GIPME). For many years, this has been operated continuously, helping revise the scientific basis on which depends our understanding of marine pollution and marine-pollution monitoring. One result has been a series of reports on *The Health of the Oceans*.[26] During the 1970s the IOC also co-ordinated research into the ocean/atmosphere interaction, under the auspices of the Global Atmospheric Research Programme (GARP) which is run by the World

Meteorological Organization and the International Council of Scientific Unions.

Undoubtedly, the key activity of the 1970s in respect of ocean pollution was UNEP's regional seas programme. This started with an attempt to persuade the Mediterranean countries that they should take co-operative action to clean up their sea. The Mediterranean is almost a case on its own: with no tide, and with its water regenerated only every eighty to one hundred years, few natural mechanisms exist to cleanse it. Yet pollution pours in from the land all around, and from the 100 million tourists who visit the region every year. The situation became so bad that in the 1970s some of the world's most famous beaches on the northern Mediterranean had to be closed because they became so polluted.

Thanks to the UNEP initiative, the Mediterranean Action Plan—or Blue Plan, as it became known—was signed in 1975. The Plan provided for nations to carry out co-operative research and monitoring; to formulate an environmentally sound development plan for the region; to prepare legal agreements committing the relevant nations to action; and to set up the necessary institutions and finance. Remarkably, the Plan was so successful that even mutually hostile nations such as Israel and Syria, Egypt and Libya, and Greece and Turkey sat down at the same table to discuss the issues. One result has been that eighty-four laboratories from sixteen countries in the region are now co-operating in a research and monitoring programme. The Mediterranean is being cleaned up.

So successful was the Action Plan that it became a model for other areas. Similar agreements have now been signed for a number of other enclosed or semi-enclosed seas: the Red Sea, the Kuwait region, the West and Central African coastal seas, the Caribbean, the East Asian seas, the South-east Pacific seas and the South-west Pacific seas. Plans exist for agreements over East African seas and the South-west Atlantic.

Reports on mid-ocean pollution during the 1970s were on the whole encouraging. While many coastlines are polluted, particularly by heavy metals such as mercury and cadmium (and by oil), the deep ocean, by and large, is not. Further studies are needed and more detailed analysis of the chemical and physical constitution of the deep ocean during the 1980s should provide a more complete picture. Studies of oil pollution during the 1970s also revealed that oil spills have not so far had permanent effects on fish populations or on fish catches—though they have killed hundreds of thousands of marine birds and temporarily disrupted related activities such as oyster farming.

In spite of this, the world fishery situation is not encouraging. During the 1970s, the world fish catch, though initially rising, levelled off—at about 70 million tonnes a year. A number of fisheries—such as anchovy in the south Atlantic and herring in the north Atlantic—decreased dramatically. And even though the gross tonnage of the total catch did not actually fall, its quality deteriorated and it became more difficult to obtain. During the late 1970s, the world fishery catch comprised increasingly large fractions of low quality, non-vertebrate, industrial fish; and fishermen had to travel increasing distances to catch even that.

According to the FAO, most fish stocks are now fully or excessively exploited. Yet it is now clear that bad fishery management, rather than naturally declining resources, is really to blame. 'It is a sobering thought', claims the major UNEP report on environmental action during the decade (*The World Environment 1972–82*), 'that in 1980 [world fisheries landings] may well have been 15 to 20 million tonnes less than they would have been had management been more competent.'[27]

The issue of ocean management is in fact becoming a key issue in almost every aspect of marine science and productivity. The Law of the Sea Conference, which had been debating since 1973 the issue of ocean resources and how they should be divided between countries, produced an international law of the sea which was to be signed and ratified during the early 1980s. It proclaims formally what is already apparent to nearly all maritime countries: the days of free enterprise in the ocean are over. As one report put it: 'The era of freedom of the seas is becoming an era of the managed ocean.'[28]

This kind of ocean management has already been exercised in preserving, among other things, stocks of herring in the Atlantic and, of course, by the International Whaling Commission which set whaling quotas for many whale species each year during the 1970s. During the 1980s whaling will probably finally become an archaic industry—though if whales are to flourish in the future, stocks may actually have to be managed rather than simply left to look after themselves.

With intensive ocean management, some scientists believe, it may eventually be possible to increase the fish catch by two or three times. This seems optimistic, in view of the fact that total global wet-fish production is currently estimated at 240 million tonnes a year.[29] However, there are undoubtedly new types of stock to be investigated—such as squid and krill—but their take will have to be regulated strictly in accordance with the needs of other forms of marine life. Whales, for example, feed intensively on krill and enough must be left to preserve the already decreased whale populations. Even so, krill may be yielding as much as 10 million tonnes a year by 1990, according to one estimate.

Some ocean resources—oil, gas, gravel, sand, magnesium, and sulphur—have been exploited regularly in large quantities for many years. But for more than two decades the minerals which lie on the ocean floor—and below it—have become more and more attractive as the costs of land-based minerals increase and underwater technology improves. As well as the famous manganese nodules, deep-water ocean drilling has now revealed sediments rich in other minerals, such as iron, copper, and zinc. How soon these will now be exploited remains unclear.

In some ways, initial enthusiasm has ebbed, partly as a result of discussions during the Law of the Sea Conference as to who should own mid-ocean underwater resources: private companies have been deterred from making large investments by the unknown legal complications which could follow any attempt to extract minerals from the ocean floor. It seems inevitable, though, that these minerals will eventually be extracted as both the technology for doing so and the international legal regime which will govern them become clarified. Interestingly, this is an area where the legal situation is likely to be resolved before the commercial technology arrives, in contrast to so many other areas of science and technology where social implications have to be dealt with *post facto*.

At the same time, there has been a resurgence of interest in the oceans as an energy source. The possibilities are many. Of the known and exploited energy sources, tidal energy is already operational—with a large tidal generating station working at La Rance, France, and more, smaller ones in China and the Soviet Union. Experiments are in progress to determine the usefulness of temperature differences which exist in the tropics between surface waters and water a few hundred metres down—OTECs, or Ocean Thermal Energy Converters, may prove to be ideal power sources for island communities near the tropics. There are also possibilities of exploiting the salinity of the sea, the currents found in it, the waves that travel across it, and the winds that blow over it.

Put together, these possibilities are substantial. According to Walter R. Schmitt,[30] they can be summarized as in Table B. By comparison, world installed electrical generating capacity in 1970 was about 10^{12} watts, and a demand of around $3 \cdot 10^{13}$ watts is projected for the next century. By then, no doubt, the attractions of harnessing ocean energy—without disturbing the human environment on dry land—will seem even more compelling than they do now.

Finally, there is need for more research on the role of the oceans in shaping other forces on this planet. The crucial link here, the interface between the oceans and the atmosphere, has been under investigation for many years but little enough is still known about it. Its importance, however, is growing. For one

Table B: The oceans as sources of power (watts)

Power source	Natural potential	Practical potential
currents	$6 \cdot 10^{11}$	$3 \cdot 10^{10}$
tides	$3 \cdot 10^{12}$	$3 \cdot 10^{10}$
waves	$3 \cdot 10^{12}$	$3 \cdot 10^{11}$
salinity	$3 \cdot 10^{13}$	$3 \cdot 10^{11}$
temperature	$4 \cdot 10^{13}$	$1 \cdot 10^{11}$
winds	$9 \cdot 10^{13}$	$5 \cdot 10^{11}$

thing, the oceans act as a carbon sink, absorbing some of the carbon dioxide which circulates in the atmosphere. If we knew the exact details of how this carbon sink works, it would be much easier to calculate the possible effects on the world climate of burning fossil fuel.

It is already clear that the oceans play a fundamental role in determining climate. During the next two decades, as agriculture presses on marginal land, as coal becomes a preferred fuel, and as several different processes threaten to alter climate, we shall need to know a great deal more about the 'buffering' effect of the oceans on climatological variations. Until we do, we cannot predict the ultimate results of human action on the planet.

In 1982 the IOC made a determined effort, as part of LEPOR, to forecast 'the aspirations and expectations of ocean scientists for research during the next few decades'. A report was produced, 'Ocean Science for the Year 2000',[31] under the organization of SCOR, the Scientific Committee on Oceanic Research. It outlines the major challenges awaiting oceanographers in the four areas of physical oceanography, the chemistry of the oceans, life in the ocean, and the ocean floor. The tone of the report is set by the final paragraph of the foreword:

To use the ocean wisely, you must first understand it. How to maintain our renewable biological resources? How to conserve the genetic potential of the 180,000 marine species known to man? How to conserve ecological integrity? What is the ultimate compatibility of the oceans for the many different kinds of pollutants transported from land by river and wind? All of these and many more problems need more and better research because at least since Bacon we have learned that 'Nature, to be commanded, must be obeyed'.

The atmosphere

The detailed constitution of the Earth's atmosphere, and the stratosphere above it, are now well known, So, too, are the general conditions governing the planet's heat balance. In theory, the incoming solar radiation balances with the outgoing infra-red radiation, via a series of complicated conversions, to provide a stable climate. In practice, there have always been minor climatic variations over the short term—and large ones over periods of tens of thousands of years. Until recently the former have been of minor importance to day-to-day life on the planet.

These variations have suddenly become significant and, consequently, have come to dominate the study of the atmosphere. There are two reasons. First, as growing populations press harder and harder on increasingly marginal agricultural land, small disturbances in climate produce greatly exaggerated effects on human populations. The Sahel drought of the late 1960s and early 1970s produced such a devastating effect for just that reason. Secondly, the human population is beginning to exert significant effects on the Earth's natural systems, and hence on the climate itself. To sum it up: 'Humanity is not only vulnerable to climate, but climate now appears to be vulnerable to humanity.'[32]

The two main threats to climate now appear to be from carbon-dioxide heating through the greenhouse effect and from destruction of the ozone layer by the use on earth of certain chemicals, most notably the chlorofluorohydrocarbons used as propellants in aerosol sprays and in refrigeration equipment. In neither case are our models of the atmosphere, or of the larger system which controls the climate, adequate to provide the necessarily exact predictions which are increasingly urgently required.

The climate itself, though determined mainly by what goes on in the atmosphere, is also crucially dependent on other factors: the oceans transport huge amounts of heat via seasonal currents; the ice-caps melt and refreeze; water vapour evaporates from the seas, the lakes, and the rivers to form cloud cover; volcanic eruptions and forest clearing add dust to the atmosphere; and the burning of fossil fuels and wood adds carbon to both the atmosphere and the oceans. A successful model of climate must take all these, and many more, factors into account and still be able to predict the effects of relatively small changes in just one of the hundreds of factors involved.

The carbon-dioxide concentration in the atmosphere has been intensively studied over the past two decades. It is known that it was 250–90 parts per million (ppm) in 1825; and that it reached 313 ppm by 1958 and 330 ppm by 1978. If consumption of fossil fuels continues to rise at the current rate of 4 per cent per annum, the level will reach 380 ppm by the year 2000, a rise of some 30 per cent over pre-industrial levels.

This rise is probably mostly due to the burning of fossil fuel, which, over the past 125 years, has released 140,000 million tonnes of carbon into the atmosphere, about half of which has stayed there. Currently, about 4,500 million tonnes of carbon are released into the atmosphere annually. To this must be added a factor resulting from global deforestation because nearly all the wood which is felled during forest clearing is burnt. At the moment, models of the process are not accurate enough to predict how significant deforestation is as a cause of carbon-dioxide buildup compared with fossil-fuel burning.

The effect of carbon dioxide in the atmosphere is to reflect back to Earth outgoing infra-red radiation. This, in essence, is the cause of the 'greenhouse' effect which threatens to increase the Earth's temperature. Currently, none of the many mathematical models of the greenhouse effect are sufficiently accurate to provide exact predictions of how much the Earth's temperature will rise as a result. Most anticipate a rise of 1.5°C to 3°C for a doubling of the carbon-dioxide level. But there are complications. Increasing dust levels in the atmosphere may counteract the process by filtering out more incoming solar radiation. A warmer climate could lead to an increase in cloud cover, causing a drop in temperature, thus counterbalancing the greenhouse effect.

The issue is currently being tackled on an international scale, thanks to the World Climate Programme launched by the World Meteorological Office in 1979. The situation, though, is potentially extremely serious, mainly because we are engaged in a process which will almost certainly affect the Earth's heat balance before we can even predict the outcome of our actions. As one commentator put it: 'Humanity has unwittingly embarked upon a dangerous experiment on a planetary scale.'[33]

During the 1980s the World Climate Programme will need expanding, and more studies must be made not only to predict the effects of increasing carbon dioxide, but to predict what results various possible changes in climate might have on human populations in different parts of the world.

These changes, of course, may not be all negative. This is not the case, however, with the other principal agent of climatic change, the effect of chemicals on the ozone layer. This layer, found generally about 25 kilometres above the Earth's surface, protects the surface from the effects of the Sun's ultraviolet radiation. Should it be damaged, the effects of increased ultraviolet radiation on the Earth below will include faster ageing of skin, increased rates of skin cancer, more eye disease, and slower rates of growth for many plants and marine life forms which live near the surface.

Once again, models of the effects of the chemicals concerned on the ozone layer are not sufficiently accurate to be able to make confident predictions. Most of the models suggest that a 1 per cent reduction in the ozone layer could increase ultraviolet radiation on the Earth by 1.6 to 3.0 per cent. If the two main chlorofluorohydrocarbons continue to be used at the present rate, the ozone layer may eventually be reduced by 5 to 10 per cent.

However, this is not likely to happen because, for once, international action has been swift. While it is difficult to curb the scale of global fossil-fuel burning, it is much easier—though still difficult—to persuade nations to control the rate of use of certain chemicals, particularly when their function in aerosol sprays can be duplicated by other less harmful chemicals. Many countries have already agreed to reduce their production and use of the chemicals, and some have already done so. The immediate objective for the 1980s is a convention regulating their use on an international scale.

So far, no one has been able to detect any change in the ozone layer as a result of the use of chemicals on Earth. If an international convention is signed in the next few years, it will be an important environmental 'first'—the first time humanity has banned the use of something which is environmentally harmful before it has actually begun to do damage. The only other possible examples of this are the various international treaties banning the use of biological-warfare agents.[34]

There is one other issue likely to play an important role in the atmospheric sciences during the 1980s: acid rain. Acid rain is caused when sulphur and nitrogen oxides, produced by power-stations, industrial plants, and motor vehicles, are washed out of the atmosphere and fall as dilute nitric and sulphuric acids in precipitation (rain or snow). Over the past two or three decades, smoke-stacks have been built increasingly high, in an effort to avoid local pollution. One result has been that the pollution is now distributed internationally by the atmosphere. Between the 1950s and the early 1970s atmospheric sulphur levels have increased by 50 per cent in Europe generally, but doubled in some Scandinavian and Central European countries. The effects include:

- corrosion of railway lines in some areas of Poland to such an extent that trains are limited to 40 km/h
- the killing of all fish in the lakes of a 13,000 km^2 area in southern Norway
- the killing of life in 20,000 of Sweden's 100,000 lakes
- prospects of damaging more than 48,000 lakes in Canada in the near future
- severe damage to forests in many parts of Central Europe
- huge cost in damaged stonework and paintwork in many industrialized countries.

The acid-rain issue—though actually one of the reasons why Sweden suggested the UN Conference on the Human Environment in 1972—has come to the fore only recently. Information about forest damage is new and alarming. And the general scale of acid-rain damage is only now beginning to be appreciated. There are, of course, effective cures in the form of filtering out the relevant oxides at the smoke-stack, but they are expensive—and the international nature of the problem tends to detract from their urgency as a national priority. Scientists will have to devote considerable time in the 1980s to studying the mechanisms by which acid rain is transported, how it can be prevented, and assessing the effects.

One problem of relevance not only here but in every other field connected with the Earth's natural systems is that of data. Satellite sensing, of course, has provided a new dimension to the study of the planet. But it has not provided substitutes for badly needed observational data on all the main parameters, with a good geographical spread, of all the natural systems. This was one of the specific objectives of the recommendations of the Stockholm conference in 1972. Yet a major study commissioned by UNEP itself had to report ten years later that 'The world community has not yet achieved one of the major goals of the Stockholm conference—the compilation, through a global programme of monitoring, research and evaluation, of an authoritative picture of the state of the world environment.'[35]

Painting that authoritative picture must now be the main objective of the 1980s.

7 Technologies

The Buddha, the Godhead, resides quite as comfortably in the circuits of a digital computer or the gears of a cycle transmission as he does at the top of a mountain or in the petals of a flower.

Robert Pirsig, 1974

Technology took on a new significance during the 1970s, as the problems of development, the environment, and energy became progressively more critical. During the 1950s and 1960s the public associated the world of science and technology more with science than with technology: the reverse is now true.

At the international level, the nature of technology has been dissected with some care. Technology is now universally accepted as a value-laden commodity, which carries with it its own concealed messages of what is good and what is bad for society. Out of this debate has risen the concept of appropriate technology, and hence by implication the recognition that technologies can be inappropriate. In development circles the issue of the right of access to appropriate technology by developing countries has become an important part of plans for the New International Economic Order. At the beginning of the 1980s we enter an era in which technology is no longer regarded as an absolute good; but, paradoxically perhaps, the importance of technology has come to be appreciated perhaps as never before.

The public image of technology has changed accordingly. Previously, value-judgements about technology extended to all technologies, which were judged collectively as good, bad, or indifferent. Today a more discerning attitude allows that nuclear energy may well be a profound environmental and social mistake while the advent of the microchip has provided the computer industry with a popularity it has certainly never before attained.

Little more than a decade ago the computer was thought of as huge and expensive; it was associated either with very large businesses or with scientists. Today the computer is associated with video games, personal computing, word processing in small offices, and the running of very small businesses. It is rapidly acquiring an image as benign now as it was malevolent a decade ago.

This chapter deals with the state of the art of technology, and of the problems and promises which exist for the future. It is not all-inclusive but instead selects four topics which seem to symbolize current technological issues: information, biotechnology, energy, and materials. The first two of these are fast-growing fields—areas in which the pace of new advance and of new social applications is as fast as any the planet has yet witnessed.

Energy, by contrast, is the field in which technical experts have for decades been predicting that scientific advance would solve all known problems for all time for all people. Just how wide of the mark that prediction was is only now becoming apparent: not only are we short of energy but some areas of the world are disastrously so. To make matters worse, no one is now confident that any kind of socially acceptable technology can solve the problem.

Materials technology is progressing in a similar way. Whereas twenty years ago materials technologists were throwing up quite unexpected new materials with astonishing rapidity, they often had little idea of their social utility. Today, the pace of advance in materials technology has slowed considerably; but very real needs have arisen for new kinds of materials, particularly those associated with energy saving, resource conservation, and environmental protection. The key question here is whether social need is a sufficient condition to guarantee progress.

Information

Since about 1950 the world of computing and information handling has undergone a revolution so profound that it is difficult to bring it into sharp focus. During the past three decades the number of bits of information that can be stored on unit surface area has increased by 10^{5-6} times and the speed at which information can be processed has increased by almost the same amount. Yet, at the same time, the quantities of raw materials and energy needed to achieve this increase in speed and capacity have actually decreased by 10^{4-5} times; and the unit cost of an electronic device has decreased by 10^{3-4} times.

These facts can be illustrated more dramatically. The number of logical/mathematical operations which it would once have taken a man his whole life to perform can now be done by a single microprocessor in just one minute. Whereas in the 1950s a computer with the information processing capacity of the human brain would have occupied the physical size of a large city, such a computer would now be no bigger than a brain. Cost comparisons are even more dramatic. If costs in the car industry had declined over the past thirty years as rapidly as they have in the electronics industry, the most expensive car would now cost about $1.

The speed at which information technology advanced during the 1970s was due to the silicon chip. Between 1971 and 1978 the number of bits of information which could be housed on a single chip rose from 1,000 to 64,000; during the same period the cost of the chip stayed constant. Within another ten years, the million-bit chip is likely to be in widespread use. Costs seem likely to continue to fall, and the pace of advance to continue unabated.

This massive cost reduction in information storage and processing has brought the chip into use in many places where it had not previously been available: in the home computer, the small office, the car engine, at the paying exits to parking facilities, at cash withdrawal points at banks, at airline booking-offices, and in the provision of raw data via the television to homes which subscribe to a service. All this seems likely to be greatly extended in the very near future. As the National Science Foundation comments in its report on the next five years: 'The coming period will see the spread of computing power to more people than ever before.'[1]

Three decades ago futurologists talked confidently of the coming nuclear age. Their optimism has proved misplaced and nuclear technology now occupies a hotly debated and somewhat ambivalent position in the list of future priorities. But, as often happens in technical advance, the nuclear age has been replaced in short order by the information age. In the next decade, the word processor will enter nearly every major office in both the developed and the developing world. In the place of the piece of rectangular paper, on which information has been stored for more than two millenniums, will come the plastic disc, a few centimetres in diameter, storing the information content of a small book.

Unlike many other technologies, microcomputing depends for its effectiveness essentially on decentralization. The microprocessor has brought computing power to the individual desk and into the individual living-room. Several manufacturers now even offer genuinely portable microcomputers—some of them of considerable power. As the technology continues to develop —with computers becoming yet faster and smaller, and being able to accept input in the form of either the human voice or visual patterns—this trend will continue. Every human activity in which data handling is involved will be affected. The implications of this are so huge that few have yet attempted concrete social forecasts of the impact of microprocessors.

The one area where advance is sure to be extraordinarily rapid is the interface between communications and computing—a field which Anthony G. Oettinger of Harvard University has called 'compunications technology'.[2] France is already looking to the time—only a few years ahead—when every telephone subscriber will have a computer terminal in his home. As fibre optics and satellite communiation become more and more common, two-way access to data banks, the ability to shop and pay for goods without leaving the home, and the possibility of communicating with other individuals via systems of electronic mail will be quickly translated from fantasy to reality.

Already, however, it is clear that some new information technologies have not developed in the ways originally predicted. Twenty years ago scientists were preoccupied with concepts of 'artificial intelligence'; computers that wrote poetry, composed music, and manufactured art were talked of in glowing terms. The comparison between the computer and the brain has not stood the test of time. The creativity of the human being has not been threatened by electronics, as most humanists knew it never would be, and the computer is now seen more simply as an efficient and fast means of handling data—a task which was hardly ever creative in human terms. Its disappearance from the vocabulary of human employment will be regretted only to the extent that it creates unemployment.

There are, however, still bottle-necks to the advance of the microprocessor. One of these is the development of the software, which has not progressed with the same rapidity as the hardware. Programming has turned out to be one of the most complex human tasks ever undertaken, and those with the knowledge and patience to do it have been richly rewarded by society. One of the strangest phenomena of the computer explosion is the ability of the young to come rapidly and easily to terms with the chip. As many experts have pointed out, the computer revolution will really occur only when the current generation of computer-literate youngsters reaches positions of social responsibility.

It is not difficult to forecast some obvious social implications of the computer age. Intelligent and appropriately trained people will find themselves once more in great demand while those without the necessary training and ability will face increasing unemployment. One estimate is that by 1990 more than 50 per cent of the work-force in industrialized countries will be employed directly or indirectly in the information business. This is already the case in the United States.

At the same time ease of communication will lessen the need for travel and personal encounter. The home will become more attractive as new types of information and entertainment become directly accessible from it. Interestingly, nearly all these advances can be made without environmental costs—indeed, many will involve environmental gains because the new information technologies consume almost negligible quantities of energy and natural resources, in contrast to newspaper technology (the Sunday edition of the *New York Times* consumes 25 hectares of forest per issue).

A French technologist, André Danzin, believes that the nature of the coming information society will be as different from the industrial society as that was from its predecessor, agricultural society. He writes:

Imagine two average Frenchmen, one born in 1900, the other in 1980. If they are 12–16 years old, the first is working, the second is at school. If they are both working, there is more than a 50 : 50 chance that the first will be on the farm and the second in an office. The amount of free time they will have during their lives is respectively 5 and 19 years. If we ask how they will spend their free time, the first will spend, outside his work, at most two years informing himself, through reading or communication, while the second will spend more than 12 years in front of his television, on the telephone, listening to the radio or at his computer terminal. These are not the same men: work has changed; the way time is used has changed; education has changed; use of leisure time has changed, radically. The most important commodity has become information and not, as official statistics might have us now believe, energy, primary materials or machines. The centre of gravity of human activity has been shifted towards the immaterial.[3]

Danzin has summarized these profound changes, which are already affecting industrial society, in Table C.

It is only when information is considered as a resource along with other resources, such as materials and energy, that the scope of what is about to happen to society can be fully appreciated. Many writers have pointed out that the information revolution now taking off is comparable in magnitude to the Industrial Revolution which occurred two centuries earlier. As Oettinger argues, action of any kind

depends on materials and energy. Without materials there is nothing and without energy nothing happens. But without information nothing has meaning: materials are formless, motion is aimless.[4]

The importance of information is highlighted symbolically by the fact that in any country where a new group seizes power, almost the first thing it does is to take over the radio and TV stations. At least, that was the situation until recently. But as Oettinger points out—and here he and Danzin appear to differ in their interpretations—these are no longer the key institutions that they were; in Iran in 1979 the new group in power did not bother with the mass-communications industry of radio and TV. Instead it monopolized the direct-dial telephone system, made extensive use of courier and mail, and flooded the country with cheap audio cassettes. There could be no more convincing proof that the

Table C: Major characteristics of three types of society

	Agricultural society	Industrial society	Information society
importance of research	minor 0.3% GNP or less	strong 1–2% GNP	very strong 2–3% GNP
importance of education	minor 1% GNP or less	strong 2–4% GNP	very strong 6–8% GNP
importance of communications	unknown	3–5% GNP	about 10% GNP
proportion of work-force in agriculture in industry in information	up to 40% 15–25% 12%	10–20% 40% or more 15–20%	10% or less 20% or less 50% or more
use of time life expectancy free time on transport	36 years 3 years 1.8 years	60–70 years 12 years 4–6 years	70 years plus 19 years diminishing
level of urbanization	up to 25%	70%	diminishing
role of mass media	minor	strong	enormous
level of organization	simple	complex	extremely complex
degree of interdependence	weak	strong	very strong
population growth	high	low	very low

Source: Danzin, André, 'Entreprise et évolution'. In *Connaissance politique*, No. 1, Feb. 1983, 116.

centre of gravity of the information industry is beginning to move away from the mass-communications industries which dominated the scene during the 1960s and 1970s.

While such forms of social evolution are of critical relevance to the already industrialized countries, they are of much less interest—so far—to developing countries still struggling to build up an industrial base. However, the importance of information as a resource raises all the familiar issues which have long been fought over other resources: who has it, who needs it, how can it be obtained, and how much should it cost? These concerns are at the heart of the developing countries' determination to establish a New International Information and Communication Order. They will be the centre of lively debates over the coming decades.

There will be plenty of other problems, too. The unemployment issue is already creating union/management friction in large companies in industrialized countries. In time, similar effects will be felt in the more advanced of the developing countries. These problems will come at an awkward moment in the evolution of human society, which is already suffering from massive unemployment caused by quite different economic forces.

The other major potential social conflict concerns the effects of the computer age on individual privacy. This issue, which has been hotly debated in advanced countries for more than a decade, shows no prospect of disappearing. On the contrary, early fears which were often dismissed as alarmist have proved astonishingly accurate. Society is likely to have a hard job during the rest of the century in reconciling its new-found powers of data handling with the possibilities of 'Big Brother' behaviour which will inevitably accompany it.

Biotechnology

The growing importance of information technology heralds an era in which 'soft' processes replace hard ones—physical travel, for example, will be progressively replaced by electronic communication. A parallel process is emerging in other fields, notably engineering, where the biological sciences are beginning to manifest their enormous power—a power which has been latent for the past century but which now promises changes as profound as those resulting from 'computronics', though it will take longer for them to take effect.

The new field of biotechnology[5] encompasses fields as diverse as genetic engineering, biological prosthetics, new fermentation processes, and the use of biological techniques to improve agricultural yields, process waste, and provide energy from biomass. The techniques involved stem from disparate areas of biological investigation. The fact that they appear to be reaching maturity at roughly the same time is not, however, accidental. To a large extent each area feeds on the next and the blossoming of applied biology which will occur over the next few decades is due to important and interconnected advances in many fields.

The most dramatic area concerns the use of recombinant DNA. The possibilities of genetic engineering have been opened up by discoveries in two areas. First, during the late 1970s techniques were developed for the rapid analysis of complicated biological molecules. By 1978 a computer 'protein atlas' had been written, containing the amino-acid sequence of more than 500 proteins. Soon afterwards it became possible for one researcher to determine the sequences of 200 amino-acids in a single working day and to determine the sequence of 1,000 of the nucleotides which form DNA in a week.

After analysis came synthesis. Although the first gene was synthesized in 1964, the technique was improved and during 1978 and 1979 it became possible partially to synthesize the genes responsible for the production of insulin and the human growth hormone; and in the latter year these genes were successfully introduced into bacterial cells. This work was made possible by the discovery of enzymes capable of cutting DNA at precise sites, of enzymes capable of sealing off the loose ends of DNA, and of other enzymes which synthesize DNA from messenger RNA.

The pace of advance became extraordinary. While in 1979 it took two years to synthesize a gene with 120 nucleotides, by 1981 the same job could be done by machine in three days. The number of genetic messages being deciphered at that time increased by 15 per cent a month, leading to a quintupling of the number of decoded genes every year. By the end of 1980 a computer gene atlas had been written, containing the sequences of 350,000 genes.

The implications of this work were that biological molecules could be synthesized to order by implanting specific genes in various types of micro-organisms. The first chemicals to be thus synthesized were insulin, somatostatin, and the human growth hormone. Human interferon—a biological compound of potential importance in the treatment of cancer—was synthesized shortly afterwards. Production of chemicals potentially suitable as vaccines followed almost immediately—first a protein theoretically suitable as a vaccine against the hepatitis B virus, and then a protein characteristic of one of the foot-and-mouth viruses.

All these products are potentially of great medical importance. The fact that they are synthesized within a cell means that they can be obtained in essentially pure form and hence are free from other toxic products. Production of therapeutic chemicals in this way suggests that long and costly toxicity tests for new chemicals may in the future be greatly reduced. Similarly, genetic engineering opens up the possibility of new methods of medical diagnosis, in which small quantities of the DNA or RNA of an invading virus are isolated, mapped, sequenced, and replicated.

The agricultural possibilities of such techniques are almost as exciting. For example, it may become possible to transfer the nitrogen-fixing genes of certain bacteria to plants such as cereals which are unable to fix nitrogen. Should this prove possible, the savings in terms of fertilizer and improved soil fertility will be enormous. Similarly, there is the prospect of transferring to a number of different crops specific genes responsible for improved yield or pest resistance. Again, the potential savings would be immense.

Another field of application concerns the tailoring of certain yeasts and other micro-organisms to do specific jobs in industry—for example, turning the by-products of oil refineries into single-cell protein suitable for an animal feed. The carbohydrate-rich waste from sugar-refining plants may also be used in the same way. It is conceivable that similar techniques may be used to produce new ranges of compounds, such as biological polymers, solvents, and detergents, capable of substituting for petroleum-derived products.

Finally, biotechnology is already finding more and more uses in the energy field. Biomass sources, such as sugar-cane, are being increasingly used as a source of ethanol for motor fuel, and agricultural wastes are being turned into biogas, with benefits both to agriculture and the rural household which thus finds itself in possession of a new fuel and an improved agricultural fertilizer.

Exactly when most of the other expected advances are likely to materialize is harder to say. Scientists themselves are currently reluctant to commit themselves to specific predictions. However a Delphi study—in which experts repeatedly refine their own predictions as they are informed of the predictions of their own colleagues—has shown that many experts are optimistic; in the figures in Table D the first date is the one at which experts predict success with a 50 per cent probability, and the second with a 90 per cent probability. These dates, of course, represent the years in which the advance is made; the time taken to reach the market-place will be longer.

In one respect the new field of biotechnology differs substantially from that of computronics: it offers immediate and attractive prospects for developing countries. Such things as the production of bio-energy, fermentation, the treatment of wastes, particularly agricultural ones, and improved vaccine production could well find application in developing countries as quickly as in developed ones.

However, as in so many areas of science, biotechnology research is largely concentrated in the advanced countries. International and regional co-operation is badly needed if the results of the powerful new techniques becoming available to biologists are to be used to improve the human condition on a global basis, rather than in the places where the research frontiers are being pushed forward most avidly.

Finally, mention should be made of some of the more curious effects which rapid, potentially commercial discoveries are having on the world of biology. Many of the teams working in genetic engineering lead a 'double life' in the sense that they work both

Table D: Delphi predictions of biological breakthroughs

Breakthrough	50 per cent probability date	90 per cent probability date
new nitrogen-fixing plants	1985	1995
single-cell edible protein	1982	1987
plants resistant to predators	1990	2000
bacteria for use in waste treatment and pollution control	1984	1990
petrochemical substitutes	1988	1995
gene therapy for diseases such as sickle-cell anaemia	1993	2010
genetic screening to isolate genes responsible for birth defects	1985	1990
mapping of human genetic code	1984	1985
better knowledge of senescence	1990	2000
understanding of immunological processes	1984	1991

for an academic institution and for a commercial enterprise—often one which they themselves have set up. Perhaps for the first time in science, the researchers are pushing back commercial and academic frontiers simultaneously. These goals tend to conflict. For example, should these scientists follow the scientific norm and publish results as quickly as possible, or follow the commercial norm of maintaining secrecy until patent registrations have been filed?

To complicate matters still further, there is some reason to believe that patent protection in this area may not turn out to be as effective as in areas where patents deal with totally inanimate objects. If so, jealously guarded research and rapid marketing may become more common even than patent registration and certainly than publication in the formal scientific press.

On the other hand, as we have seen elsewhere in this book, the formal distinction between academic research and applied science has been beginning to break down for some time. In this sense what is happening in biotechnology is not that unusual but is symptomatic of trends which can be found throughout the sciences. The dilemma may be more apparent in biotechnology than in other fields only because the rate of advance there is currently so high.

Energy

No field contrasts with those of information and biotechnology more dramatically than that of energy. While the former are undergoing the kind of explosive growth which many people came to associate with science and technology in the 1960s, energy is encountering all the problems which the increasingly hostile public of the 1970s predicted for it.

While the problem certainly started with the oil crisis of the 1970s, which has directly or indirectly quintupled petroleum prices since it began, it is a common misconception that all problems stem from the OPEC countries. The fallacy of this argument is easy to demonstrate. Were the supply of oil the only problem, alternative solutions could have been produced. Nuclear power, intensively developed since the early 1950s, had reached a kind of 'take-off' position just at the time of the oil crisis. Yet nuclear power did not blossom in the wake of the oil crisis. On the contrary, it withered. Whereas in the United States there were in 1973 no less than thirty-four new orders for nuclear reactors, there were only two in 1978. By 1980 the nuclear power industry was virtually at a standstill in many of the developed countries.

Clearly, much more is amiss in the energy field than an uncertain and costly supply of oil. Even allowing for the fact that nuclear power cannot provide all the substitutes we need for a liquid fuel such as oil, the stagnation of the nuclear industry needs more detailed investigation. The role which public opinion has played in limiting the nuclear options is examined in Part III

(Chapter 9, 'The public and the media'). Some of the other issues include:

- heightened awareness of the environmental implications of many big-technology energy projects
- growing appreciation that the planet has a finite size and that therefore not even science can necessarily produce unlimited abundance
- a suspicion that 'growth solutions', often environmentally and financially expensive, are not the only solutions—a suspicion borne out by the small or even negative energy growth rates in many developed countries during the late 1970s.

Energy expenditure per capita varies widely throughout the world, paralleling income distribution. In the United States gross per capita energy expenditure is well over 10 kilowatts; in some regions of some developing countries, it is less than 100 watts—barely enough to light a single electric light bulb. The energy problems of the developed and developing regions must therefore be considered separately.

In the developed world, the two major fuels of the past decades—oil and natural gas—will approach scarcity by the end of the century. In many countries, in fact, peak oil production has already been passed and the time of peak natural gas production is not far ahead. As is now well known, neither fuel will actually run out—they will simply become scarcer and more expensive as time goes on.

The two major substitutes that are envisaged until the end of the century are coal and nuclear energy. Coal is still plentiful—world supplies are assured for a period of something like 300 years—but is a dirty fuel. It caused most of the developed countries' severe air pollution problems earlier in the century. Furthermore, real fears now exist of altering the Earth's climate by burning excessive amounts of carbon fuels, giving rise to a greenhouse heating effect in the Earth's atmosphere from accumulated carbon dioxide (see Chapter 6, 'Natural Systems'). Paradoxically, this danger has been increased partly as a result of the environmentalists' insistence on the dangers of nuclear power.

At present, though it is technically possible to convert coal into either a gas or a liquid fuel—and even to do so underground without having to mine the coal first—the process is not economic. However, the need for liquid fuels has stimulated much new research in coal technology, an area which was almost completely neglected during the decades in which confidence in oil, natural gas, and nuclear energy abounded.

Though nuclear power is now competitive—if the enormous capital costs of development are discounted—with most other methods of generating electricity, two major problems beset the industry. The first is that the nuclear-fuel cycle currently used exploits only about 0.7 per cent of the natural energy of uranium. It is thus expensive on fuel. Because uranium is not abundant on the Earth, the current nuclear-fuel cycle has only a short-term future.

However, this problem could be solved if fuel reprocessing and breeder reactors were used. The technology of both operations is considerably more complex, and all the safety issues which surround conventional nuclear power—including the still un-solved problem of disposing safely of nuclear waste—look more dubious if breeder reactors and fuel reprocessing are brought into the equation. The latter also raises the question of nuclear-weapon proliferation, for the fuel cycle used depends on the production of plutonium, the key ingredient of nuclear weapons.

So acute have the safety and non-proliferation issues become that development of breeder reactors has been halted in a number of countries. In spite of this, the breeder reactor is the only nuclear reactor currently envisaged which could have a future much after the turn of the century. As the US National Research Council has put it:

As a technology for producing electricity, nuclear fission is at least as reliable and economical as coal, its closest counterpart in the production of baseload power. But concern with its safety is persistent, and the issues of waste disposal, reactor safety, and weapons proliferation must be convincingly and publicly addressed if the technology is to have a productive future.[6]

In 1984 these issues had still not been effectively addressed.

Nuclear power remained unpopular and many countries were planning substantial increases in coal burning to provide their electrical energy.

Most were also intent on exploiting the new and renewable energy sources as far as possible—the Irish, for example, were replanting peat bogs with trees to fuel timber-burning power-stations, many countries were growing biomass crops, including potatoes, as a source of the fuel ethyl alcohol, and several were planning large experimental wind and solar power-stations. Others were experimenting with wave and tidal energy plants, and with OTECs (Ocean Thermal Energy Converters) designed to exploit the energy available in the oceans as a result of the temperature difference between the surface and deeper waters in tropical regions.

These innovations, combined with substantial efforts to conserve energy—with improved insulation, more efficient boilers, and better control systems—were combining to reduce demand for scarce fuels. Motor vehicles, for example, became dramatically more efficient in terms of fuel economy between 1975 and 1982.

Yet the fact remains that the energy sources of the future are still undetermined. Clearly, the fuel economy of the future will be a mix and all kinds of energy from tidal to biomass and from nuclear to oil-shales will have to be used. But in the long term there are only three principal choices: fusion energy, breeder reactors, and solar energy in one form or another.

A large question mark still hangs over fusion energy—the Sun's energy source, in which simple atoms are fused together, forming larger atoms and liberating huge amounts of energy. Though progress has recently been made there is now no chance of a commercial fusion power-station being in operation before 2010–20; and there is, of course, still the possibility that the technology will escape us entirely and that fusion will turn out not to be a workable solution.

Solar energy, in all its forms, is the most attractive possibility. The energy is there in sufficient quantity and with conversion to hydrogen could provide the required fuels as well as baseload power. Solar energy has one other advantage not shared by either fusion or fission technology: it is adaptable in scale. Appropriately engineered, it can be used to power anything from a massive centralized solar power-station to a minute, remote, 100 watt solar-cell installation. If the history of technology has taught us anything over the past two decades, it should certainly be that flexibility of this kind has real social value.[7] Unfortunatley, the technologies available for exploiting this diffuse source of power are still primitive. Far more research is needed.

Finally, the breeder reactor is, paradoxically, the most developed of the alternatives—the one workable solution which could be put into large-scale practice after the turn of the century. But there are major problems. It will be interesting to see whether our greed for power will prove sufficient motivation to overcome our scruples with regard to safety and weapons proliferation, or whether caution will prevail. The outcome may well depend less on technical factors than on the skills and effort which the nuclear industry devotes to its public relations—a field in which it has so far not excelled. While this may not be a very 'scientific' attitude, it is symptomatic of the democratic participation which the general public is increasingly demanding in all high-technology decisions.

The situation in the developing countries, though less often discussed, is in fact far more serious. Apart from those few developing countries which are oil exporters, the oil crisis had serious effects on many developing countries' economies. Suddenly as much as half or even three-quarters of a country's hard-earned foreign exchange had to be spent on oil imports. The inevitable result was a sharp decrease in available power and a severe shortage of foreign exchange which put many countries back into the situation from which they had just begun to emerge a decade previously.

At the same time, the combined effects of overgrazing, deforestation, and population growth were beginning to take their toll. In some developing countries, 85 per cent of the energy used is in the form of fuel-wood. For some 1,500 million people, wood, dung, and crop wastes are the only accessible energy sources. Their annual consumption is estimated at 1,000 million m^3 of wood—well over 80 per cent of developing countries' total wood use, excluding exports.[8]

But fuel-wood is now becoming increasingly scarce. In the Gambia it now takes 360 woman-days a year per family to gather the sticks necessary for cooking and heating.[9] An FAO survey[10] has shown that already 100 million people have insufficient fuel-wood to meet their needs and another 1,000 million suffer from fuel-wood shortages. By the end of the century, it is estimated, 2,300 million will be in desperate need of an immediate alternative to fuel-wood to meet their energy needs.

Yet the UN Conference on New and Renewable Energy Sources, held in Nairobi in 1981, concluded that there was no realistic alternative to fuel-wood in view. Progress can come only through such measures as reforestry, use of wood-burning stoves, improved wood economy, and the use of wind power, biogas, minihydropower, and other alternatives wherever possible.

One area of possible hope is photovoltaic conversion. Many developing countries are blessed with prodigious amounts of sunlight but village electrification, long held to be essential to rural development, seems unlikely to be accomplished through the conventional grid system, due to the enormous capital cost of linking an estimated 1,000 million remote villages world-wide to their respective national grids.

If the price of solar cells could be substantially reduced, however, each village could run a small photovoltaic unit, supplying power for light to each home and running perhaps a communal water pump and TV set. Solar-cell costs are falling—and they could be made to fall much more rapidly if large orders for solar cells were placed with manufacturers. But we meet another chicken-and-egg problem. The price of the cells is too high for large orders to be placed, and the expected cost reductions are not occurring. International co-operation is urgently needed to bring down the price of solar cells, and provide a desperately needed alternative energy source for village life.

Centralized, commercial power sources in the developing countries are also a problem. Nuclear power is currently both too expensive and unpopular. The developing countries cannot opt for the developed countries' temporary alternative of coal-fired power-stations, for, with few exceptions, coal supplies are very limited in the developing countries. However, there are extensive hydropower reserves in the developing countries (only 9 per cent of the hydropower potential of the Third World has so far been exploited) and these are providing some respite in some of the richer areas such as Brazil. Minihydropower units are also becoming popular, following China's example; in the past twenty years China has installed more than 88,000 minihydropower plants, with an average capacity of 70 kilowatts.

Globally, energy has become the world's number-one resource problem. Attempts to solve it via conventional high science and technology means have not been successful and are not now likely to be. Instead, the world is being pushed towards a plurality of sources, and research is being directed at each. The challenge to the world's scientists and technologists which is resulting from this is enormous but the research has not so far been efficiently organized on an international basis. Here is an area of knowledge transfer where organizations such as Unesco, UNIDO, and ICSU could make enormous contributions.

Materials

Materials science and technology made dramatic progress between the 1950s and the 1970s. During that time many new materials arrived on the market, of which the two most notable were a new range of plastics and the micro-electronic materials which made the contemporary computer revolution possible.

Revolutionary advances of this kind are no longer being made, and indeed are not expected over the next two decades. There are, however, a number of focuses of research where results are expected to influence society profoundly. At the same time there are a number of areas where social needs cannot be met by current technology—for instance, there are needs for less energy-intensive methods of materials production, for easier means of recycling primary materials, and for useful materials which can be made from non-petroleum bases.

Future advances in the field must stem from one of two sources: either the state of knowledge in materials science will make possible new materials with new uses, or social demand will pull innovation in such a direction as to stimulate action in certain required fields.

By and large, it is the latter force which is currently making the running—indeed, a constantly recurring theme of this book is that throughout much of science and technology the pull of social demand is now proving more of a stimulus and catalyst than the push of advancing science. As the US National Research Council report on the next five years comments: 'In the main the initial impetus in the materials science and engineering process seems to arise more often from "technological pull" than from "scientific push".'[11]

It is not difficult to list some of the major materials needs of the kind of society which is emerging in the 1980s. They include:

- high-temperature superconducting materials, badly needed in fusion research. The highest critical temperature at which a material is superconducting is currently 23 K. High-temperature superconductors would save huge amounts of energy and other materials such as conducting metals.

- single-crystal alloy components, needed primarily in turbines and other high-temperature machinery. Production of single-crystal components eliminates 'grain boundaries', producing materials with improved characteristics. Such materials are likely to become available shortly.

- corrosion- and oxidation-resistant coatings and materials. There has been some success—note the long life of jet aircraft engines, due mainly to improved nickel-base alloys, new coatings, and elements made from carbon fibres. But corrosion resistance has not yet been significantly extended to the motor vehicle and other products.

- glassy metals are about to become available. Subjected to very fast cooling, these new materials are metals which resemble glass in their amorphous structure. They are strong and have excellent magnetic properties. They could be used in transformer cores, and in the United States alone it is estimated that in the power transmission industry they could save the equivalent of 6 million barrels of oil a year.

- new dielectric materials to replace the polychlorinated biphenyls (PCBs) are badly needed, as PCBs are now widely regarded as environmentally unacceptable. (There are no real signs of substitutes as yet.)

- cheaper means of producing solar cells (see 'Energy', above) are badly needed. The price breakthrough seems likely to come not from new materials but from sheer volume of production.

- non-oil-based plastics. While the technology exists for producing many types of polymers from fossil forms of carbon other than oil, the prospect of producing them from renewable resources is much more remote (but see 'Biotechnology' above).

- development of the silicon-carbide and nitride ceramics, which do not occur naturally, is one of the few materials fields which could soon provide dramatic new products. They may be used to replace metals in high-temperature machinery and even as a basis for a non-metallic car engine.

- new materials sensitive to infra-red radiation and the effects of pressure are likely to become available, making it possible to 'see' in the dark and by sound waves. New information technologies of this type could have unforeseen social consequences and do not necessarily respond to any social need.

These are some of the major interests of materials technologists today. As has been stressed, they do not necessarily correspond to the current promises of materials science, and for this reason progress may be patchy. As in the field of energy, social problems are proving more important as the motivation for innovation—in contrast to the two fields of information and biotechnology where new advances in the state of the art are what are making the running.

The study of current trends in technology thus provides two examples of fields in which dramatic advances are being made, much in the manner of the 'big-science' style of the 1950s and 1960s. In the other two fields—energy and materials—the pace of development is less dramatic but the social need for solutions very real indeed.

There are thus two distinct problems for policy-makers. The first, and familiar one, is that of controlling the explosive growth of a rapidly advancing field to ensure that the social consequences are fruitful and benign. The second, less familiar, problem is to try to catalyse the growth of research in certain key areas where social demand for innovation is pressing.

Part III · Choices

The world we see before us in the mid-1980s differs so markedly from previous experience that it is hardly surprising that the way ahead looks problematic. The problems, however, are tactical rather than strategic. As we examine the role of the principal actors involved in shaping the relationship of science and technology to society over the coming years, we must thus constantly bear in mind three facts:

- the future is likely to depend more, rather than less, on science and technology than it has in the past
- the problems which face us are more urgent than they have ever been before
- these problems are essentially world-wide in nature and cannot be resolved by either one set of actors, such as the scientific community, or by one nation or group of nations. They require global solutions.

During the past decade two entirely distinct types of study have reinforced each other's conclusions. Concern with the environment and energy supplies has vividly demonstrated that our problems are global ones, shared by every nation in the world. At the same time those who examined the basic tenets of the development process during the 1970s[1] reached an identical conclusion: problems in the developed world cannot be solved without reference to developing countries, and the latter cannot develop without the co-operation of the industrialized nations.

In the past, experts on the global science and technology system have often concluded that there were two main problems: one concerned the role of science in society in industrial nations; the other was to stimulate the endogenous development of science and technology in developing countries. While these problems can still be separately formulated, it seems doubtful that they can be separately resolved. They are, in fact, both part of the much broader problem of harnessing science and technology to the all-important task of providing a decent way of life for every human being on the planet Earth over the coming decades.

In looking at the choices which lie ahead, we shall analyse the role which can be played by each of several actors: the scientific community, the public and the media, the nation state, the transnational enterprise, and finally the international community.

Throughout, one basic problem will emerge repeatedly. Integration of the science and technology system with the political process—whether nationally, regionally, or internationally—is difficult. One reason is that the two things operate on different time-scales. As Ibrahim Helmi Abdel Rahman has pointed out,[2] the time-scale on which the science and technology system operates is typically of the order of thirty years. A national government rarely survives much longer than five years. Hence at the national level, at least, mechanisms have yet to be found for impressing on short-lived governments the necessity of long-term strategy in science and technology.

The converse is also true, though more rarely discussed: a science and technology which could respond more quickly to political need would be of great value. As Unesco's Director-General, Amadou-Mahtar M'Bow, told the UN Conference on Science and Technology for Development in August 1979: 'The developing countries are facing an unprecedented challenge, that of mastering science and technology, in the shortest possible time, in order to ensure their own survival.'[3]

This problem of time-scale is at its most severe at the level of the national government. Regional and international institutions, as well as national and transnational enterprises, tend to have longer lives than national governments, and hence are able to mesh more easily with the natural rhythm of scientific progress. This conclusion adds further weight to the global emphasis currently being applied to development issues.

The time-scale involved is not, however, by any means the only reason why integrating science with politics is so difficult. The other reasons concern the nature of science and the nature of the communities of which it is composed.

8 The scientific community

In other words, scientific research is a social activity. Technology, Art and Religion are perhaps possible for Robinson Crusoe, but Law and Science are not.

John Ziman, 1968

Scientists work in many places. They are to be found in universities and schools, in almost every kind of industry, in government research organizations, and indeed very often in government itself. On average no more than one-third of them are actively involved in research though nearly all of them do at least some teaching. They sit on committees, they write papers for learned societies, and they are almost invariably enthusiastic advocates of the virtues not only of their own profession but also of their own speciality within it.

In what sense, then, does this amorphous group of people form a community? And, if it does, of what relevance is this to the way science is conducted, in both the developed and developing countries?

The reason why scientists group themselves into communities is primarily to do with communication. All forms of modern scholarship grow like trees, with new fields of investigation constantly branching out from the main trunk of learning which has been in existence for decades or even centuries. Research is possible only if it builds on existing knowledge. To do research necessitates being familiar with what others are doing. In the seventeenth century this could be achieved simply by reading the occasional book published by the occasional scholar. Today, as John Ziman has put it, 'A scientific laboratory without a library is like a decorticated cat'.[1]

By the 1950s scientific publishing had become a major industry, and the research paper the key measure of individual performance in science. Such was the speed of growth in science that by the late 1960s formal publication in the scientific literature was being augmented by a system of 'pre-prints' in which the scientist would distribute advance copies of his work to selected members of his specialized community even before it was formally published. It was this group which Derek de Solla Price immortalized with the name 'The Invisible College'.[2]

In this chapter we shall discuss the three forms in which the scientific community exists. The first of these forms is the invisible college. The second we shall call 'the visible college'—the system of societies, associations, journals, unions, and clubs which have been expressly created to cater for the needs of scientists. The third form is more political, involving the structure of government organizations through which science is managed, financed, and directed. We shall call it 'the political college'.

Clearly, there is overlap between these three forms. Individual scientists will play different roles in all three of the systems, hopping from a ten-minute chat with fellow scientists in their invisible college, to attending meetings of a scientific society in a visible college, and then sitting on a committee to advise on future financing for their particular fields. The important point, however, is that each of the three systems is essential to the health of scientific life. Politicians are normally aware of only the last of the three—though they do give grudging approval to the existence of the second. They understand virtually nothing of the first. Yet it is the first which constitutes the very stuff of science.

The invisible college

So far in this book we have said little of how science actually works. The process is, in fact, very different to the popular image of what science is and how it progresses. The absent-minded professor beavering away on his own, producing one work of outstanding genius late on in his lifetime, is about as distant from reality as the image of the mad scientist trying to perfect the ultimate doomsday machine. John Ziman puts it thus:

The scientist is seen as an individual, pursuing a somewhat one-sided dialogue with taciturn Nature. He observes phenomena, notices regularities, arrives at generalizations, deduces consequences and eventually, Hey Presto, a Law of Nature springs into being. But it is not like that at all. The scientific enterprise is corporate. It is not merely, in Newton's incomparable phrase, that one stands on the shoulders of giants, and hence can see a little farther. Every scientist sees through his own eyes—and also through the eyes of his predecessors and colleagues.'[3]

But if every scientist must see through the eyes of his colleagues, he must have access to them. That access is obtained through communication—a communication which, to be sure, is achieved partly through the scientific literature. But only partly. It is achieved through discussion, by letter writing, by phone calls, by attending conferences, by reading the circulars of the invisible college. An outsider watching a scientist at work will be amazed to discover that the scientist spends most of his time apparently engaged in idle chatter, drinking cups of coffee. But this is the hub of his work. The gathering of information and the exchange of ideas are the essential preconditions for research.

Formal channels exist for the exchange of these ideas—the scientific conference, the learned paper, the evening discourse. But these are no longer the keys to the process. Any scientist worth his salt will know in advance exactly what is going to be said at next week's conference or what will be described in the next issue of his discipline's learned journal. He attends the conference not to hear the papers but to talk to other scientists about their latest ideas and interests—picking out of the air the ill-defined substance of tomorrow's theories.

The reason why this process is so important is that science is cumulative, each step building upon the last to create an edifice of knowledge to which most of the scientists in any particular field have made some contribution. It follows that, however well equipped a laboratory may be, it will never produce important work in isolation. Another way of saying the same thing is that a scientist shipwrecked on a desert island can never do science.

There are, in fact, two reasons for this. The cumulative effect just mentioned is one. The other is that the results of science are in the end established by consensus. The scientific community exerts a powerful filtering effect on the products of its work. It insists that any published research must contain sufficient detail to enable the experiment to be repeated, and hence confirmed or denied, by other scientists anywhere in the world. Papers which are published must pass through a complicated refereeing system before they appear in print. Research proposals go through a complicated system of committees before being given the scientific go-ahead.

In all these processes, the age, sex, skill, character, and colour of the scientist concerned is of no relevance. In the invisible college all are equal. The twenty-year-old Ph.D. student has equal stature to the Emeritus Professor. This is the world of ideas. And while on occasion there may be abuses, there may be victimization, there may be greed and competition—as in any human activity—the norm and the ideal are clearly established and in theory at least universally respected in the world of science.

It follows that the idea of secrecy is simply incompatible with scientific research. Science, as John Ziman has said so eloquently, is 'public knowledge'. If that knowledge is not public, it is not science because it cannot play any part in constructing the edifice of theory on which research scientists must build.

The invisible college is the shadowy structure which scientists have created to make their craft work. It defies complete description yet is as essential to their work as the test-tube and the scales. It is also one of the major reasons why developing countries find such enormous difficulty in making their own scientific communities flourish. Poor as they are, often geographically remote from other centres of research, and unaccustomed to the activities of a breed of men who do indeed behave rather differently from others, the conditions under which science can flourish are too often denied. If a scientist cannot play his allotted role in his invisible college, he might as well return to farming, banking, or merchandising—or whatever. Certainly, his career as a scientist is at an end.

A word of caution, however, is in order here. The Western model of the scientific community is something that may never fit happily into the different social, cultural, and economic conditions which prevail in many developing countries. As much is evident from the lack of success of so many attempts to import the model into the Third World. Unfortunately, this model is almost the only known working version which is familiar to the world scientific community. Though there are certainly other ways of arranging the interface of science and society, and ways which could be of great benefit to developing nations, they have yet to be discovered.

The visible college

In the early days of science, the invisible and the visible colleges of the scientific community were one and the same. All started with the idea of the scientific society in the seventeenth century. Among the earliest were the Italian Accademia dei Lincei (1603), the British Royal Society (1662), the French Academy of Sciences (1666), and the Berlin Academy of Sciences (1700). By 1790, it is estimated, there were no less than 220 of them.[4]

They functioned as invisible colleges, as meeting-places where scientists could get together to discuss their ideas. Somewhat later they also invented the idea of the formal publication of the results of scientific research in what was to become the learned journal. Later still, they also began to attract research funds, and to employ research scientists. At this point they began to combine the three roles of invisible, visible, and political colleges.

Today the scientific societies still perform a little of all three functions but their main job is as the spearhead of the visible college.

In a developed country the scientific community's visible college is a complex organization. There is a club for the élite—the Royal Society in the United Kingdom, the National Academy of Sciences in the United States. There are societies or associations for each particular branch of science. There is an association concerned with the promotion of science, such as the American Association for the Advancement of Science. There is a prestige journal, such as *Science* or *Nature*, and the increasingly common popular scientific publications such as *Scientific*

American and *New Scientist*. There are specialist journals for each discipline.

Very often there will also be unions not only for the scientists themselves but also for technicians. There will also be the countless national committees of regional and international scientific associations. Most of them will also have *ad hoc* committees dealing with individual disciplines.

These visible institutions are formidable in their complexity. Yet their functions are not hard to find. They do two jobs. The obvious one is that they act as pressure groups on society. It is through them that scientists make their voices heard. Through them they lobby for more funds, new laboratories, and improved salaries. And because of this they give the scientific community its identity. The individual scientist is enabled to feel a part of a larger enterprise, an enterprise indeed which plays one of the most important roles in contemporary society.

However, the visible college is not formally part of the political structure. It is outside it—and being outside it, it can play the most important role of government critic and adviser. Independence, in this sense, does not necessarily mean financial independence. Many of the institutions of the visible college are supported, at least in part, by government funding. But this need not, and indeed must not, compromise the intellectual independence on which any scientific community must insist.

In the eyes of many people in Europe and the United States, science's visible college has of late become too closely associated with government. Certainly it is true that much of the visible college of science has an establishment image. For this reason there has grown up a tradition of scientific association, in which scientists join together to press for social reform of one kind or another.

One of the oldest and best known of these is an international association known as the Pugwash Conferences on Science and World Affairs, a group comprised mainly of distinguished physicists whose major concern is to prevent the outbreak of nuclear war and slow down the crippling arms race in which the major powers are now involved. More recently, Pugwash has also been turning its attention to problems of development and energy supply.

The rationale behind organizations such as Pugwash is that their role as scientists allows them a freedom of speech, and an ease of communication, which is not readily available to politicians. During a Pugwash conference, the scientists present speak as scientists and not as representatives of the countries to which they belong. This has indeed enabled them to bridge political gaps which politicians would find far too wide for comfort. It is important to note, too, that associations of this kind believe that the special expertise they have aquired as scientists enables them to speak with more authority on some issues than can other groups in the general population.

As scientists have become increasingly concerned about the technologies to which their research has ultimately given rise, yet broader kinds of groupings have occurred. Their role is subtly different. They believe not that they can pronounce with more authority on certain issues—more democratically, they tend to believe that the views of all sectors of the population should carry equal weight—but that as scientists they have a greater *responsibility* to examine the implications of the social changes to which science and technology give rise. The British Society for Social Responsibility in Science is one of the best known of this type of association.

However, there is one factor common to all these newer types of association: they have come into being as a result of a desire to inject an ethical dimension into the conduct of science, and to introduce the idea that scientists should feel social responsibility for the effects of their work. The idea is not new to science, even if the associations are. In *New Atlantis*, Francis Bacon wrote three centuries ago that:

This we do also: we have consultations, which of the inventions and experiences which we have discovered shall be published and which not; and take all an oath of secrecy, for the concealing of those which we think fit to keep secret: though some of those we do reveal sometimes to the State, and some not.[5]

One of the fundamental reasons why developing countries find it hard to establish endogenous science and technology is that no framework exists in those countries on which to build the scientific community's visible college. Even in the more advanced of the developing countries, scientists find it extremely hard to establish a national identity: in countries such as Argentina and Brazil there are still few equivalents of the major science institutions of the industrialized nations. And in the poorer countries scientists have to work in almost complete isolation from one another, and from their colleagues in other countries. In such conditions, science is unlikely to flourish. It is even less likely to produce the cultural and economic rewards which are the *raison d'être* of the scientific community.

Once again, however, we must beware of making too facile a comparison between the forms of scientific community which have evolved in the industrialized nations and those which might be best suited to a developing country. The fact that the former appear not to work well in developing countries may simply mean that the model is wrong.

The important point here is that the creation of endogenous science means much more than establishing and financing research laboratories. The scientific community resembles a beehive. A colony of bees and a hive are two ingredients. The third, and most often forgotten one, is the social organization which goes with it—in the case of a beehive, a social organization of enormous complexity which alone can ensure that the hive functions efficiently, in effect as a single organism. A scientific community is no less demanding, and no less dependent on its own organization.

The political college

In one sense science is in a unique position with regard to politics. Traditionally, at least, it is the only field which advises government on how to spend money on itself. This somewhat curious situation has arisen because scientists claim that only they are in a position to assess research priorities. Consequently they have a say in how large the national research budget should be, and a major say in which particular areas it is best spent. They also defend, sometimes extremely fiercely, their right to manage research themselves, without the help of external bureaucrats. As Ziman has stressed:

To put, say, an admiral in charge of a laboratory—even an admiral with a Ph.D.—might be as offensive and stupid as putting a professor (even a professor with a master's ticket) in command of a battleship. Whether or not professional scientists should be given executive power in general government, there is no doubt that the management of scientific research must be in the hands of those who are respected for their scientific and intellectual achievements.[6]

The system by which science is financed and managed varies a great deal from country to country. A number of scientists will actually be employed by government, and charged specifically with this responsibility in various fields. In the American system, there is one key adviser, the scientific adviser to the President, and each government department has committees with special responsibility for scientific research. In the European systems, there is more generally a government department for science (sometimes coupled with education) which has overall responsibility for financing and managing all research. In either case, the necessary advice comes from committees of scientists formed for that purpose.

The relationship between a scientific community and the body politic is extraordinarily subtle in the industrialized world. In many ways it defies accurate description. As Jerome B. Wiesner, the scientific adviser to President Kennedy, once said: 'Without really having planned it, in fact I would say without really understanding it, we have evolved a system, as we often do in this country, of checks and balances and interactions that gives us the strongest scientific and technological community you can find in the world.'[7]

The details of these systems need not concern us here. What does concern us, however, is the nature of the broad interplay which exists between science and politics. The temptation, of course, in any new developing country has been to place the emerging scientific community firmly under the political wing. Yet a scientific community cannot flourish without a degree of independence. One of its key roles is to advise government. This it cannot do if it is a part of government. Writing nearly twenty years ago Don K. Price concluded his book *The Scientific Estate* with these words:

Perhaps indeed a nation can be free only if it is not in too great a hurry to become perfect. It can then defend its freedom by keeping the institutions established for the discovery of truth and those for the exercise of political power independent of each other. But independence should not mean isolation. Only if a nation can induce scientists to play an active role in government, and politicians to take a sympathetic interest in science (or at least in scientific institutions) can it enlarge its range of positive freedom, and renew its confidence that science can contribute progressively to the welfare of mankind.[8]

Such a lesson is not easily learned. As we shall see, even the making of a national science policy is a complicated and difficult matter. The business of setting up a network of advice and control in science is even more difficult. The systems which appear to work in the developed countries have many critics, are not easy to describe or replicate, and are in any case in a state of rapid change.

This means that models can rarely be borrowed from existing set-ups and satisfactorily adapted and improved for work elsewhere. The one requirement which is easily spelled out is far more difficult to apply in practice: the government must learn to trust its scientists. If it does, it can expect advice from them which in the long term will stand the country and its scientific community in good stead. But that advice will bring no great rewards in the short term and provide no easy solutions to the pressing problems of development. Science is not a magic wand; it is, on the other hand, an essential ingredient of sound development.

In one sense, it may be lucky that the science advisory systems of the developed world cannot be simply imported into developing countries. Those systems are based, without exception, on a model of science, technology, and application which may be both outdated and irrelevant to a developing country. In one way or another, they all subscribe to the view that basic research should be supported both because it provides new knowledge and because that knowledge may also turn out to be useful. Applied research is supported in order to turn potentially useful knowledge into practical application.

Yet, at the beginning of the 1980s, it is clear that in some fields at least the world of science and technology no longer works in quite that way. In biotechnology, at least, the frontiers of knowledge are being pushed back at precisely the same time as new applications emerge. The process is one, and cannot be divided into pure and applied fields, any more than it can now in environmental science. It may well be that we are entering a new era, and one of which we know little. There is, in fact, now an urgent need for detailed investigation of the dynamics of the science/technology interaction in these new fields. Much may be learned about a relationship of crucial importance to the future of both science and society.

For developing countries, caught as they are between the urgent need to solve practical problems and the apparently less immediate need to build up endogenous science, this could be a happy accident indeed. Perhaps developing-world science will find a way of dispensing with the long lead times traditionally associated with science and technology, and of creating a form of investigation which satisfies simultaneously the need to seek for truth and the need to provide practical solutions to specific questions.

Gerald Holton was clearly aware of this when he told a Unesco meeting:

We can now see many signs that young scientists and engineers are ready for a different vision for themselves, a vision of a *double service*, both for truth and for mankind's urgent practical problems . . . Unesco seems to me the only candidate today through which to articulate and animate this vision for double service of our creative scientists and engineers.[9]

The ways in which science reacts with politics, and vice versa, are changing in other respects as well. As we have seen, scientists defend their right to manage their own affairs, and even to advise governments on how much should be spent on science, and which science, with some vigour. But their entreaties to do this are regarded less favourably now than they were. The social control of science, it is increasingly argued, is indeed a proper matter for politics itself, and the special pleadings of scientists should carry no more weight than, say, the special pleadings of architects or plumbers or historians or solicitors.

During the 1980s this process is due to be taken several steps further. As the political process is subjected to increasing democratization, there is a growing feeling that major technological decisions, and even scientific ones, should be taken not by scientists, not even by politicians, but by the public at large. As we shall now see, important precedents have already been established. What scientists do in their laboratories, it is being argued, must be accounted for not only to the body politic but also the population at large.

9 The public and the media

This is the era of 'public participation' in science. The once widespread feeling that scientists alone should have domain over the scientific enterprise is being replaced by a philosophy that calls for public involvement in science, irrespective of the fact that many of the elders of science find the idea abhorrent.

Barbara J. Culliton, 1979

Until the early 1970s the general public's role in science and technology decision-making was nil. In fact, in nearly every country in the world the public's role in any kind of decision-making was very small. In the democracies, elected representatives are, in theory, meant to solicit the views of their electorates, and vote on major issues accordingly. In practice, of course, these representatives must usually follow the party line. Even in democracies, the free vote is a rare thing.

Public participation in decision-making became a major issue only in the early 1970s. And public participation in science and technology decision-making is still a new phenomenon. But although it may still be resisted both by governments and by the scientific community, it has undoubtedly come to stay. The public voice has already affected the planning of major roads and new airports in countries in many different areas of the world. It has stilled the pace of nuclear energy expansion in a dozen countries. And in the United States, at least, it has also had an important effect on the ways in which professional scientists conduct research, particularly in the fields of health and genetic engineering. Perhaps even more significantly, the public lobby against nuclear weapons was, in the early 1980s, becoming ever more vociferous and demanding. It was beginning to become a major force in the buildup of pressure for nuclear disarmament.

If it be accepted that the public has a right to have its say in the general process of decision-making, its right to do so in the arena of science and technology is more controversial. Traditionally, this has been regarded as a field in which the issues involved are so complex that only experts can have a legitimate opinion. In practice, this concern with expert opinion has in the past often been used as an excuse to exclude people from participation in science and technology decision-making. The latter, after all, has always been done by politicians, however well briefed they may have been by their technical advisers. But if politicians can be informed sufficiently well to exercise legitimate authority in the science and technology arena, then so can the public.

There are, in fact, a number of reasons for arguing that public participation in science and technology decision-making is even more justified than in decision-making in general:

- science and technology decisions may affect society more rapidly than others
- many of the issues raised, such as weather control, are entirely new
- many are orders of magnitude more complex than other issues
- some have irreversible, global implications
- many threaten to undermine deeply held social values
- the public is often remarkably suspicious of high technology.

When issues as important as these are at stake, it is clear that mechanisms have to be found for greater public participation. The search for such mechanisms dates from the early 1970s when a number of European countries first became concerned about their nuclear futures.

Nuclear controversy

The first requirement of any decision-maker is access to sufficient information. As a recent Australian report emphasized, 'No matter how skilled an individual may be, how sophisticated the techniques at his disposal, or how perceptive his judgment, a lack of adequate information will seriously prejudice him and impair the effectiveness of any decision he has to make.'[1]

In a few countries, such as the United States, freedom-of-information acts have been passed which enable any individual to obtain information from the government unless it endangers national security. Although there were many initial fears that this kind of legislation would simply mean that civil servants spent most of their time processing information requests from the public, this has turned out not to be the case. Where such acts have been passed, it has been found that governments are not swamped by requests from the public; and, secondly, that in fact it is journalists rather than the public itself who have made most use of the acts. Countries contemplating acts of this kind can now do so with one worry removed.

However, even freedom of information is clearly insufficient for a public intent on participation. Few individuals have either the time or the knowledge to conduct long personal inquiries into subjects of special interest to them. Hence public participation requires action by government to open up decision-making processes, and to help get information to people who need or want it.

Sweden was one of the first countries to attempt this, late in 1973. Its motivation was to improve understanding of the issues affecting Sweden's nuclear future, and its technique was the mechanism of study circles. Public opposition to the nuclear power programme was large and growing; the government feared that it might not even have sufficient backing to carry its programme through Parliament.

The experiment lasted for a year and in that time seven institutions ran 10,000 energy study circles in which some 80,000 people participated. The government provided documentation for each circle and publicized the scheme under the banner: 'Learn more, and you will have more influence. Join an energy study circle.' The total cost of the project was about $650,000.

Four public hearings on energy issues followed and in March 1975 the government put forward a more modest nuclear energy programme than it had originally envisaged. There had clearly been feedback from the public to government, and government had acted accordingly. One result was that the bill got through Parliament. However, the change of government that followed in

September 1976 was at least partly due to the strong anti-nuclear stand taken by the party which came to power.

In spring 1978 Sweden made plans to spend $183m. on renewable energy—mainly biomass and wind—and on conservation measures. This was a new programme and the decision to put Sweden's seventh nuclear power plant into operation was left pending.

The overall results of the study-circle programme are somewhat ambiguous. First reports indicated that the programme polarized debate and did not reduce public uncertainty. However, when the programme was stopped, resistance to nuclear energy was markedly high, which partially explains the change of government which followed. Subsequently the Swedes' attitudes to nuclear energy appear to have mellowed, as the results of an opinion poll show (Table E). The question asked was 'In a referendum, would you vote for or against nuclear energy?'

The study-circle idea may be a particularly Swedish approach to the problem. A number of other countries—notably Austria, Denmark, and the German Federal Republic—have tried the different but more obvious technique of mounting public information programmes.

In the Federal Republic of Germany the process was known as the Burgerdialog, or dialogue with the citizens. It began in 1975 with advertisements in newspapers and periodicals with a coupon for citizens to fill in to obtain technical information on nuclear energy. All told, a million copies of three publications on nuclear

Table E: Results of Swedish opinion poll on nuclear energy

	For	Against	Don't know
October 1976	27	57	17
May 1977	32	49	19
September 1977	35	46	19
March 1978	39	40	21
September 1978	41	37	22

Source: Sifo AB, Vallingby, Sweden.

energy were distributed, and the programme cost $300,000 in the first year.

During 1976 and 1977 some thirty-seven seminars and discussion groups were organized and a set of information packages sent out to schools and adult education centres. This part of the programme cost about $1.4m. a year. The following year the programme was again expanded, specifically to include employer and trade-union organizations; the cost mounted still further, to $3.2m.

The expense, however, did not prevent the rise of one of the most vociferous anti-nuclear movements in Europe. Major public demonstrations took place in both 1976 and 1977 against plans for two new nuclear reactors, and in both cases government-approved construction plans were turned down in separate court decisions. Later in the same year a similar fate befell two more reactor plans.

The government also studied the problems of waste disposal and fuel reprocessing. Plans were made to site these facilities in Lower Saxony, where salt domes exist of the right type for the deep underground storage of wastes. However, the state government of Lower Saxony then objected; indeed, it went further and ordered an international critical review of the problem by a panel consisting of ten US experts, four UK ones, four from Sweden, and two from France. The panel did not recommend building the new facilities and in 1979 the decision was postponed—perhaps the first time in Europe that a national decision has ever been questioned and rejected by a team of foreign experts.

In Austria an even more curious history was to befall nuclear power. The decision to build Austria's first nuclear power plant was taken in 1968 and public reaction was not aroused until the mid-1970s when plans were announced for two more plants. A public information programme was begun but was substantially disrupted by the anti-nuclear lobby. A series of meetings was organized to debate the nuclear issue, and some of them were televised. At some of these meetings the anti-nuclear faction was in the majority, and took advantage of this to vote in a new chairman and debate a different agenda from the one originally planned.

The Chancellor then decided on a national referendum. To quote the OECD:

On November 5, 1978, the Austrians voted against government plans to put into operation Austria's first nuclear power plant at Zwentendorf. Although the referendum was to have an advisory and not a binding effect on Parliament's decision, the Austrian Parliament subsequently voted not to commission the nuclear plant, thus effectively terminating Austria's nuclear power programme.[2]

Similar public information campaigns have, of course, been undertaken in many other countries including Canada, Denmark, France, the United Kingdom, and the United States. In the latter the anti-nuclear movement of the 1970s produced substantial political impact, halting the construction of dozens of nuclear power plants and sending the industry into a state of economic decline from which it took a very long while to recover.

Two questions need to be asked about these public information programmes. First, to what extent did they achieve their objectives? And, second, to what extent did they increase public participation in decision-making about nuclear-energy options?

Several observers believe that the public information programmes did little to alter people's views on the desirability of nuclear energy, although they certainly did fan the flames of public debate. In Sweden the results of opinion polls seemed to indicate a swing towards nuclear power after the study-circle programme ended, but in other countries public information programmes appeared to increase anti-nuclear feeling. In Denmark opposition to nuclear power grew quickly when the information campaigns were concluded.

The important point, however, is that these campaigns were a poor substitute for public participation in decision-making. In many countries the public did indeed participate in nuclear decision-making—but not in the way their governments intended. The participation usually took the form of massive public demonstrations. Sometimes the very size of these demonstrations was sufficient to indicate to governments that the strength of anti-nuclear feeling ran stronger than they had suspected, and they adjusted their nuclear programmes accordingly.

The public programmes were really something of a red herring. As the OECD comments:

the main motivations behind decisions to initiate public information campaigns appear to have been to defuse controversy, gain time and thus avoid having to take quick decisions that might have politically divisive repercussions.[3]

Even this interpretation may be a little generous. Most of the European campaigns, with the possible exception of Sweden's, look more like rather naïve attempts on the part of government to sway public opinion in their favour. Those who have studied the area tend to see the information put out more as propaganda than as anything else. For instance, Dorothy Nelkin, of Cornell University, analysed the situation in three European countries and concluded that 'the experiments to date surely represent more an effort to convince the public of the acceptability of government decisions than any real transfer of power'.[4] Nelkin admits, however, that the experiments did have the result of making some governments modify their nuclear programmes, and they also apparently increased public demand for a greater say in other areas of science and technology policy-making.

The role of the media

Traditionally, of course, there has always been one indirect route of communication between government and the public—via the media. To be sure, the messages transmitted may not always be exact and they may not always be of the kind which governments prefer. However, even the public information campaigns which have been conducted relied on the media to a considerable extent.

The professional science writer has benefited greatly from the public's increased interest in science and technology decision-making. This has given him, in effect, increased leverage on his editor and hence more space in his paper, and more time on his radio or TV station. In the past, straight science stories were always a hard sell in the world of journalism. Now that they are spiced with the heady perfume of controversy, placing the story has become easier. And as it becomes easier, so too can more information be included—the effect is reinforcing. The reporting of science and technology in the mass media improved immeasurably, in both quantity and quality, during the 1970s.

Realizing this, many countries have undertaken active programmes to improve science writing and reporting—often as part of a public information campaign. One study in the United States has shown that, surprisingly perhaps, the public still receives most of its information on science and technology from newspapers and magazines. Television, apparently, is less useful and the more formal educational channels much less so.

In fact, of course, the science press has always been a highly useful means of getting information across to the public. Governments planning information campaigns might, indeed, be well advised simply to restrict their campaign to science writers and broadcasters; for a tiny investment they could in this way reach an audience of millions at a fraction of the cost of trying to do it directly.

The role which the public and the press play in shaping science and technology policy in developing countries is somewhat different—though far from negligible. So far, of course, there are few equivalents in the developing countries of the public information programmes and study groups found in Europe and the United States. However, the media have played a major role in trying to focus science and technology in developing countries on relevant topics.

Over the past two decades, many developing countries have invested heavily in universities and research institutes in the hope of attracting some of the benefits which science and technology have apparently brought the developed countries in the past. So far, of course, this investment has not paid off in the sense that basic needs have still not been met and industrial and economic expansion has been far slower than was hoped.

The media have been quick to contrast the massive investment with lack of concrete results. In a number of developing countries, particularly in Africa, their complaints have forced governments to restructure their science policies and institutions in an attempt to make them respond more quickly to the problems of development. As in the industrial countries, such moves have often been resented by the scientific community which frequently feels its freedom is being attacked and its own wisdom questioned. Nevertheless, these actions have forced the scientific communities to re-examine their roles in society and have brought the public, the media, and the scientific communities into open debate. That debate now seems unlikely to be concluded before satisfactory accommodation has been made on all sides. In this sense, at least, the role of the media has been positive.

ABMs and SSTs

Although nuclear power was by far the most spectacular area of technological confrontation between the public and governments during the 1970s, there have been plenty of others. In the 1960s there was widespread opposition to plans both to build anti-ballistic missile systems (ABMs) and to proceed with expensive projects to develop supersonic civil airliners (SSTs). In those days, opposition was about all that was allowed; certainly there was no question of public participation in the expensive Anglo-French decision to proceed with their supersonic aircraft, Concorde, in spite of the cancellation of a similar project in the United States.

In retrospect, the public's wisdom over the SST issue was not only commendable but should give confidence to those concerned with the future of the public-participation issue. Whereas in the late 1960s it was projected that by 1985 there would be some 500 SSTs in daily flight in the stratosphere, only about a dozen Concordes were actually in regular international service in 1983. With the round trip London–Washington–London costing more than $5,000, half-empty flights were the norm rather than the exception. There was clearly no possibility of recouping the thousands of millions of dollars spent in development. Not a single aircraft had ever been sold to a foreign airline.

The SST issue also illustrates another aspect of public participation; important though adequate information is, it is not always the paramount issue. The public's reaction to supersonic flight does not need to depend on a knowledge of technical parameters to do with thrust, engine specification, or even environmental consequences. However, it does have to do with social priorities—how great is the need to reduce marginally the flying time over the major oceans? If there is any benefit, how large a proportion of the population will reap it? And how much will it cost those who will not benefit? As the OECD report on public participation in science and technology decision-making points out:

it would be too simplistic to conclude that public opposition or protest in areas of scientific and technologically-related controversy, such as nuclear energy, can be attributed solely or even mainly to insufficient public knowledge and understanding. Many of those who oppose certain scientific or technological programmes are extremely well-informed as to their details and associated risks. Nor can attitudes of public scepticism be explained by one reason alone. Often they result from a broad panoply of public concerns and misgivings about the social goals to be pursued, the protective measures to be taken, and the way costs and benefits are to be distributed in society-at-large.[5]

Questions like these frequently underlie the key technical issues of our time. Important though it is to ensure that people are kept as fully informed as possible about the technical nature of the issue, it is equally important that governments keep themselves informed of the social wishes of their peoples. Participation requires dialogue; neither propaganda nor monologue can substitute for it.

The ABM issue of the 1960s illustrates a different facet of the problem. Though it may now be accepted in principle that the public should have a say in decision-making about major technological projects, that acceptance does not include decisions of strategic importance. No government has yet asked its electorate to decide whether or not it should acquire nuclear or other weapons of mass destruction. No government has even held a referendum on it. There may be debate; there may even be a certain amount of information-pushing about the technical parameters involved. But participation is another matter.

During the late 1960s the two major nuclear powers began to

consider the possibility of protecting their key cities with ABM systems, at huge cost. One result of this was to spur on the arms race, for attempts were then made to design new types of missile and warhead which would be able to penetrate an ABM system. In the United Kingdom it was considered essential for the theory of deterrence that the UK Polaris missile could strike at Moscow. An attempt was therefore made to produce a new generation of warhead which could penetrate the Soviet ABM system known as Galosh.

In fact, the ABM systems never really got off the ground. But the new generation of warheads did. In the UK the programme was called Chevaline and it was hidden from public gaze until 1980. By that time it had cost £1,000m., a budgetary item which had never even appeared in the annual defence estimates. The relevance of this story to the debate on participation is that certainly the public knew nothing of all this; nor did the elected Members of Parliament; and nor, it now turns out, did most of the members of the British Cabinet. The decision to go ahead with Chevaline was taken, it is said, by just three or four members of a privileged inner cabinet, one of whom was the Prime Minister.[6]

Stories like this indicate how far some governments still have to go before public participation becomes a reality. Token participation in minor technical issues may establish the principle. But the point is not to establish principles but to allow people a say in decisions which may literally affect their own survival. If the public's right to have its say over matters of civil nuclear-energy policy is now grudgingly allowed, that right has yet to be extended to the field of nuclear strategy, where the role of the expert is still held to be paramount.

The Cambridge affair

In the early 1980s it seems perfectly understandable that supersonic airliners, with a dozen exceptions, should not form part of our technological life-style. Yet two decades ago such a prospect loomed up on planners and technologists as a major nightmare. To have left a technical option unexplored at that time was identified by a great many people as 'the end of human progress'; up till then, whatever was technically possible had almost invariably been developed, tested, and put to use.

The decision by the United States not to produce an SST was the first of its kind. It provoked fierce controversy. This was the first time, certainly in 300 years and possibly for millenniums, that a clear technical advance had been established as possible and then rejected. It is difficult now to recall the furore which the decision provoked. How far we have come since then is shown by the fact that today it seems utterly unremarkable that we should (almost) be without supersonic civil transport.

A precisely parallel situation has now been reached in the conduct of science, as distinct from technology. The area remains highly controversial and concerns the right of the public to play a role in shaping social rules about scientific research. The issue was first highlighted during the mid-1970s at Cambridge, Massachusetts, where the City Council threatened to take legislative action which would prevent biologists at Harvard University and the Massachusetts Institute of Technology from carrying out certain experiments. The council decided to take advice from a specially appointed panel. And that panel consisted not of scientists but of members of the general public.

The experiments concerned the use of recombinant DNA. The potential danger that arose was that biologists—as they acquired the ability to slice up the DNA molecule, make changes in it, and then recombine it with other bits of DNA—might end up producing a 'virus' which was highly virulent, either to man or other living things. The safety issues which then surfaced concerned the chances of such a virus surviving and reproducing outside the laboratory, the ease with which it could be detected if it did, and the severity and timing of any effects it might produce.

It is important to note that these dangers were first highlighted in public in 1973 by two scientists involved in work on recombinant DNA. They saw a need for responsible and concerned action. They alerted other scientists and the following year a scientific conference formally requested that the US National Academy of Sciences study the subject. It did, and it recommended a partial moratorium on two experiments which at that time were judged potentially hazardous.

With hindsight, it is easy to see that most of the drama which was created during the late 1970s over recombinant DNA was actually unnecessary. It is now widely accepted that the hazards which were postulated then were greatly exaggerated. Much of the fuss was created by scaremongering by the popular press, aided and abetted by a few distinguished environmentalists and radical scientists. However, the affair is highly relevant to this chapter because, whether the proposed experiments were hazardous or not, it has led to an important re-examination of the role which the general public should play in controlling scientific research.

In 1976 Harvard University proposed building a special high security laboratory in which to conduct its experiments with recombinant DNA. The issue raised controversy both within the university and without. The mayor of Cambridge—the city which houses Harvard—organized two public hearings on the issue and the Cambridge City Council then voted for a three-month moratorium on construction of the new laboratories.

The Council then set up a citizens' review board to investigate further. It reached its decision in January 1977; construction could go ahead, but stringent safety regulations would be imposed. Similar legislation was to follow in many other US States and communities.

The Cambridge Review Board made two remarks which have earned a place in the history of the science/society relationship. The first reads:

Knowledge, whether for its own sake or for its potential benefits to humankind, cannot serve as a justification for introducing risks to the public unless an informed citizenry is willing to accept those risks. Decisions regarding the appropriate course between the risks and benefits of potentially dangerous scientific inquiry must not be adjudicated within the inner circles of the scientific establishment.[7]

The Review Board was composed of eight citizens, all non-scientists, chosen to represent their local community. The fact that they succeeded in making any headway in such a technically complex area surprised many people. The Board commented:

We wish to express our sincere belief that a predominantly lay citizen group can face a technical scientific matter of general and deep public concern, educate itself appropriately to the task, and reach a fair decision.[8]

That, however, was not a view which found wide acceptance within the scientific community. Many biologists involved in this kind of research felt that the dangers had been greatly exaggerated and the kinds of restrictions which were being proposed were far too severe. Many of them also reacted rather sharply when they saw their freedom of action apparently being curtailed by a panel of lay people. Although similar legislation was passed in other States, and Federal legislation was also prepared, the scientists mounted considerable pressure to get legislation relaxed. They succeeded.

The backlash created was undoubtedly partly due to the fact that lay people were apparently telling scientists, in effect, what they could or could not do. While this may be a bitter pill for the scientific community to swallow, it looks as though it may have to get used to it. It is, after all, no different in principle to the long-established tradition of verdict by jury, a tradition in which lay people, with no specialist knowledge of the legal process, pronounce the accused innocent or guilty according to the evidence presented to them and its interpretation for them by a judge.

If this analogy be pushed further, it should be recalled that an important and related principle of Western justice is that the accused is held to be innocent until proved guilty.[9] This was certainly not the case in the Cambridge affair. Once the scare had been perpetrated, the biologists concerned were certainly widely judged to be guilty until such time as they were able to prove their innocence. This proved difficult for them because they found themselves unable to manipulate the press as easily as they had hoped or as quickly as they wanted.

One of the most sensible accounts of the controversy, and its implications, has been written by Barbara J. Culliton.[10] Her conclusion is one from which the scientific community can derive some comfort:

Public participation is not dangerous for the scientific enterprise. It is time-consuming, there is no doubt about that, as those who have defended clinical and basic research can well attest. And it is likely to lead to restraints that previously were not imposed. Nevertheless, the restraints that come from ethical considerations and recognition of the need for public accountability cannot be dismissed as inappropriate. To the contrary, they may lead in the end to greater public understanding of and appreciation for science. In any case, they are part of the social cost of democracy.

The principle that lay people should now be involved in regulatory activity over certain kinds of research seems fairly well established. In the United States, the Federal government now requires that 'bio-safety committees' draw 20 per cent of their members from the general public having no connection with the institution performing the research. In the United Kingdom the Genetic Manipulation Advisory Group comprises eight scientific or medical experts, five members of the general public (one of whom must be chairman), four members nominated by the Trades Union Congress, and two others representing the management of industry and of universities.

In the United States the National Commission for the Protection of Human Subjects of Biological and Behavioral Research has included members of the general public since it was established in 1974—no more than five of its eleven members are allowed to be engaged in biomedical or behavioural research. The Commission has already reported on a number of key areas, including research on prisoners, children, foetuses, and psycho-surgery. Some of these reports have been included in official guide-lines on what is permitted and what is not in the general area of human experimentation.

Questions for the future

The public's role in shaping policy connected with nuclear power and research on DNA manipulation is well documented and easily established. However, these two rather glamorous examples are but the tip of a much larger iceberg.

In the field of technology, the public has participated in the past decade in innumerable public hearings and courts of inquiry—two notable ones including the three-month-long Windscale inquiry in the United Kingdom on nuclear-fuel reprocessing and an even lengthier affair (lasting twenty months) on the Alaskan natural gas pipeline. Exercises like these are long and costly; they do, however, have the double advantage of acting as information sources for almost every kind of interested group—general public, consumer groups, environmental movements, policy-makers, and government (including local government) itself.

In the field of science, action can be anticipated in many more areas than have so far been highlighted. For example, the list of 'Concerns about science' included in Part I (in Chapter 1, 'Science for knowledge') gives a fair summary of the spread of issues with which the public can be expected to concern itself over the next two decades. So far public involvement has concentrated mainly on the areas characterized as dangerous or objectionable research. But infiltration to the other areas now seems certain to come. As Dorothy Nelkin puts it:

The negotiation is no longer over whether there will be greater public control of science, but over who will participate in establishing controls, how controls will be organized, and how much they will influence detailed decisions concerning the nature and procedures of research.[11]

There are thus two sets of questions which now need answering. The first is procedural: how can public participation in science and technology decision-making be effected? Existing techniques are cumbersome, expensive, time-consuming, and inefficient. The only real exception is the referendum, which is a swift and efficient way of testing public opinion. It is not, however, a means of involving the public in actually making decisions.

In fact, there are two schools of thought on how the procedural issue may evolve. One is that science and technology are no different from other human activities, and that public accountability can therefore be effected using well-known and well-tried existing mechanisms. As the US attorney Harold Green has said:

I do not see anything that is inherent in science that ought to distinguish it from any other aspect of our society in terms of the operation of the political process. Everything else is subject to the adversary process and debate, why not biomedicine?[12]

On the other hand, equally compelling evidence exists that science and technology are special human activities which do deserve different treatment from other fields. Public accountability itself, of course, is not likely to be questioned. But how that accountability is to be exercised remains problematic.

The second set of questions goes somewhat deeper. The French writer Paul Valéry suggested that 'all politics is based on the indifference of most of those concerned, without which politics would be impossible'. Quite possibly, Valèry is at last being proved wrong—though politics certainly looks a great deal more impossible than it used to. However, these are the very early days of public participation and our ignorance of what is likely to be involved is certainly much greater than our knowledge.

As is often the case, it may be wise to start studying not the immediate questions but those which underlie the whole issue of participation. As the OECD puts it:

it may be that the most important and long-term implications lie elsewhere; on psychological attitudes and on public perceptions, feelings and emotions. Why some people participate while others abstain, why some political systems can adapt and others do not—these are questions that cannot be avoided if one is seeking to understand the essential nature of public participation and its implications for government and for society-at-large.[13]

10 The nation state

A country without an indigenous scientific or technological capacity has no means of being aware of its own needs, nor of the opportunities existing in science and technology elsewhere, nor of the suitability of what is available for its own needs.

From the *World Plan of Action*, 1971

'The development of science in the long run', wrote Adriano Buzzati-Traverso in *The Scientific Enterprise, Today and Tomorrow*,[1] 'is incompatible with the existence of sovereign nation-states.' Yet only a few years later the Vienna Programme of Action on Science and Technology for Development was to come to almost exactly the opposite conclusion. 'The ultimate goal of science and technology', it stated in its Preamble, 'is to serve national development and to improve the well-being of humanity as a whole.'[2]

These two views represent the extremes of a spectrum of opinion. The first is the voice of science, claiming in effect that science must be left to its own devices, untouched by the authoritarian ways of the nation state—a type of human organization which, some scientists contend, has much to learn from science. The second represents the view of the national politician, intent on using the fruits of science to further national development.

In this chapter we shall start by analysing the basis of this controversy because it sheds light on the problems currently confronting developing countries in their search for endogenous capabilities in science and technology. We shall end by looking in more detail at the successes and failures of countries trying to stimulate scientific growth.

The industrialization of science

Throughout its long history in the West, science has been involved in a number of successive battles with authority. Galileo and Darwin took on the might of the established Church of their times, and emerged in the end victorious. There have been similar battles with secular authority. But certainly up to the middle of the nineteenth century, there was no formal relationship between science and the State. In England the Great Exhibition of 1851 was to mark the turning-point. Previously scientists had, as Hilary and Steven Rose have pointed out, 'the status of amateurs and gentlemen'.[3] By the 1850s scientists were beginning to call for public money to support their work.

They did not do so unanimously. The radical tradition of science, and the individualism of many scientists, combined to provide a powerful force which contested anything that smacked of planning for science. The apologists of academic freedom fought a fierce battle. Their case was weakened by the role which science was to play in the First World War—a role which was to be so important that the war itself became known as 'The Chemists' War'. And then, just as the movement for academic freedom in science was beginning to reassert itself in the 1920s and 1930s, there came the Second World War—'The Physicists' War' (later still, the war in south-east Asia was to become known as 'The Biologists' War'). From 1945 on, there was never again to be any question that science could exist outside the jurisdiction of the State. The age of scientific bureaucracy had arrived. Academic science was transformed into industrial science.

With it went a tradition of long and noble standing. The Soviet physicist P. L. Kapitza was to write:

The year that Rutherford died (1938) there disappeared forever the happy days of free scientific work which gave us such delight in our youth. Science has lost her freedom. Science has become a productive force. She has become rich but she has become enslaved and part of her is veiled in secrecy. I do not know whether Rutherford would continue nowadays to joke and laugh as he used to do.[4]

In the world of industrial science, the practices of former times are turned on their heads. Where once the great universities of the world were concerned primarily with what Sir Eric Ashby called 'individual men',[5] they are concerned in the regime of industrial science with the production of what was at first called 'scientific manpower', then 'qualified scientists or engineers' (QSEs), and finally, and even more strikingly, 'human resources'.[6]

Industrial science is also the world of statistics on research and development, of the huge multinational projects of 'big science', of ministries for science, of science and technology policy, of 'grantsmanship',[7] of science parks, and of Delphi studies of likely breakthroughs in research. None of these terms today strikes a discordant note to the well-attuned ear. Yet it is as well to remember that they are recent: none is more than forty years old, many a great deal less. They are the fruits of a revolution in thinking about science which has come to pass within a single generation and yet which often escapes our attention.

In many ways, it is irresponsible to attack the industrialization of science. In a world where injustice and scarcity play such a dominant role, it is surely contentious to pretend that the planet would benefit more from the unfettered meanderings of academic freedom. Yet it is only twenty years since Jacques Barzun made a rallying call for science simply on the grounds that it was 'glorious entertainment'.[8] It is important, however, to realize that the critics of industrialized science do have valid points to make. According to J. R. Ravetz:

The assimilation of the production of scientific results to the production of material goods can be dangerous, and indeed destructive of science itself. For producing worthwhile scientific knowledge is quite different from producing an acceptable market commodity, like soap. Scientific knowledge cannot be mass-produced by machines tended by semi-skilled labour. Research is a craft activity, of a very specialized and delicate sort.[9]

It is perhaps already academic to enquire whether the delicate fabric of the research craft has been irreversibly damaged by industrial science. The revolution has occurred, and there will be no going back. During the period 1940–80 the debate between the romantic conservatives of science and the abrasive progressives of the technocracy has been ebbing and flowing.

There is no doubt about who has won. There is doubt, however, as to where the debate may lead. Some find grounds for optimism that the sterile debate of the past two decades may be replaced by a new conception of the relationship of science to society.

Guy Gresford, who was Deputy Secretary-General of the United Nations Conference on Science and Technology for Development, is one of them. In 1978 he told a Unesco Symposium on University/Industry Interactions in Chemistry that:

The sense of global responsibility which is growing among the scientific community is a logical sequel to our better understanding of the place of science in society. Science should attend to practical problems; if it has a social conscience, it follows that science should concern itself with those in greatest need. The 1980s will, I think, mark the ascendancy of the development ethic, just as the past decade has seen the flowering of the environmental ethic.[10]

Science in national development

The key issue which national governments in developing countries have to contend with is the distribution of research scientists and research funds over the globe. In 1978 the regional distribution of these was as shown in Table F.

While these figures provide a vivid illustration of the concentration of research and development in the developed countries, they do not, of course, illuminate what has been happening in the developing world. In the ten years 1968–78, for example, the total number of scientists and engineers in the region of Asia and the Pacific increased at an annual rate of 6 per cent.[11] In a few developing countries the rate was much higher. And what has been happening in Asia has been repeated elsewhere—in Africa and Latin America. There has, in fact, been something of a university explosion, with large new campuses springing up in many parts of the world. This is one way, clearly, in which the developing countries intend to promote the endogenous growth of science and technology. Unfortunately, the results are not always quite what were intended. As one Unesco document has put it, 'Higher education is growing far more quickly than employment opportunities.'[12]

Table F: Research and development by region (1978)

	Percentage of global number of R & D scientists and engineers	Percentage of global R and D expenditure
Africa*	0.7	0.4
Arab States	1.4	0.5
Asia* **	25.6	17.0
Europe	36.6	40.0
Latin America and Caribbean	3.3	1.8
North America	31.0	39.3
Oceania**	1.4	1.0
Totals	**2,131,500**	**US$127,654m.**

  * Excluding Arab States.
  ** In Asia Japan's share was 74.7 and 89.5 per cent respectively and in Oceania the share of Australia and New Zealand was 98.3 and 99.1 per cent respectively.

All data in the tables in this chapter exclude the USSR, China, Mongolia, Vietnam, and the Democratic People's Republic of Korea, for which countries comparable data were not available.

Impressive though these figures are, they also hide a more alarming tendency. University growth in the developing world has been generally higher than 6 per cent a year. But in many countries it has proved difficult to attract sufficient numbers of students in science and engineering. The popular subjects have been the social sciences and humanities, where enrolments typically account for 50 to 70 per cent of the total (compared to 10 to 25 per cent for engineering).[13]

The reasons are not hard to find. As we shall see, in developing countries most scientists are employed by government, in the general services sector. They are often poorly paid, have little or no opportunity to do research, and do not enjoy high status. Students are well aware of this. They are also well aware of the fact that salaries in private industry may be three or four times as high, but that industry in the developing world rarely carries out research. The implication is obvious: a degree in business studies or some related topic is the most useful. Hence the high proportion of students enrolling in the social sciences.

In spite of this, the numbers of scientists and engineers working in research and development has risen sharply during the past decade. In Asia and the Pacific the developing countries' share of total research and development (in manpower terms) rose from 23.8 per cent in 1974 to 29.1 per cent in 1978. Overall, the numbers of scientists and engineers working in research and development in the developing countries rose from about 170,000 in 1974 to 250,000 in 1978—a growth rate approaching 20 per cent per annum.[14]

But this major achievement, at least in quantitative terms, needs to be seen in perspective. The numbers of scientists and engineers working in research and development per million inhabitants is one of the most revealing national statistics (Table G).

While progress in manpower has been significant, the other major indicator of research growth—percentage of GNP spent on research and development—suggests a smaller budgetary response. At the beginning of the 1970s the UN Advisory Committee on the Application of Science and Technology to

Table G: Number of R and D scientists and engineers per million inhabitants (1978)

Africa	53
Arab States	202
Asia (excluding Japan)	99
Northern America	2,736
Latin America and the Caribbean	209
Europe	1,632
Oceania (excluding Australia and New Zealand)	118
Japan	3,548
Soviet Union	5,024 .
Australia	1,617*
United States	2,685

* 1976.

Development (ACAST) proposed in its World Plan of Action that developing countries should devote at least 0.5 per cent of GNP to research and development.[15] Even that modest goal has not been achieved (Table H). In the mid-1960s the average values for developing countries of percentage of GNP spent on research and development were 0.1 to 0.3 per cent. By 1980 they had reached only 0.2 to 0.4 per cent, though figures were substantially higher in such countries as India, Argentina, Brazil, and the Republic of Korea.

These figures also hide the way in which the money is spent. In the developed countries research and development takes place in three quite distinct areas: industry, the universities, and government research laboratories. It is not uncommon to find that private industry spends almost twice as much on research and development as does the government—whose funds are divided between university research and government laboratory research. And there are incentives for university/industry co-operation. While the lack of efficient transfer mechanisms between university and industry are frequently and publicly deplored in many developed countries (other than the United States), such mechanisms do nevertheless exist. And compared to what happens in most developing countries, they are models of efficiency.

The first problem in developing countries is that private industry, by and large, does not invest in research and development. In many developing countries there is, quite simply, no private industrial research and development at all. In other

Table H: Percentage of GNP spent on R and D (1978)

Africa	0.45
Arab States	0.31
Asia (excluding Japan)	0.42
Northern America	2.15
Latin America and the Caribbean	0.49
Europe	1.88
Oceania (excluding Australia and New Zealand)	0.29

words, the state budget for research and development is always the major source of funding, and sometimes the only one. This is true even for relatively advanced developing countries such as India where, in 1979, industry funded only 14 per cent of total expenditure on research and development. Under these circumstances, there is little prospect of the engine of science and technology providing added push to economic growth. As a Unesco paper has put it:

In most developing countries, the government is still, for all practical purposes, the sole source of funding and the agency which executes most research work. This has led to the well-known situation of two worlds cut off from each other: the research systems, subject to government administrative rules, and the production system subject to the forces of international economic competition.[16]

Quality in research

The figures quoted in the previous section give only a quantitative summary of what is happening in developing countries. Because the economies of nearly all developing countries are strained, the promotion of research is subject to constant checks and cutbacks; research priorities have to be carefully selected; and the debate as to whether to support basic and applied research, or just applied research, seems never ending.

Unesco has spelt out some of the things that can go wrong:

The factors governing the productivity and effectiveness of research groups need to be carefully analysed so that the main obstacles can be identified. Examples of such obstacles are the operation of research institutes as bureaucracies; using managerial methods developed for purely administrative services; the selection of projects without first consulting industry and other research users; the lack of adequate funds; people combining several jobs because of insufficient remuneration; the lack of team spirit among researchers; and the lack of status and officially recognized careers for professional scientific researchers.[17]

The importance which should be attached to fundamental research is one of the most vexed issues. Although it is difficult to make precise definitions of what is basic or fundamental research, and what is application-oriented, it seems that on balance most industrialized nations devote between 15 and 25 per cent of their research to basic studies. The figure for Japan is nearer 35 per cent.

In some developing countries—mainly those with close, perhaps post-colonial links with a developed nation—the research effort has been surprisingly academic, and almost unrelated to national development plans, with many scientists working in the most abstract areas of mathematics and theoretical physics. While it is hard to see the relevance of such work to a country with pressing development problems, it does have one, often-ignored advantage: the scientists involved are likely to be better integrated with the world scientific community—where concentration of effort in these areas is common—than are scientists working on development issues, which are still largely ignored in the national research programmes of the developed countries.

A survey[18] by Unesco's Division of Science and Technology Policies has shown that, overall, the countries which place most emphasis on fundamental research are those which seek to play a major role in international relations. On the other hand, there are an increasing number of developing countries which, for one reason or another, have more or less abandoned any attempt at basic research. Instead their scientists concentrate almost exclusively on local problems closely related to national development. They tend to emphasize work on appropriate technologies and alternative energy sources. Such countries are normally those without major ambitions in the international arena. They tend instead to place great emphasis on national self-reliance and the preservation and development of indigenous culture. The price they pay, of course, is relative isolation from the international scientific community.

The quality of fundamental research in developing countries is difficult to assess. One rather dubious measure of the extent to which a national scientific community's work is recognized internationally is the extent to which its work is published and read by other scientists. In 1978, in the region of Asia and the Pacific, 94 per cent of the scientists who published papers in the specialized journals came from the Soviet Union, Japan, India, and Australia. New Zealand provided a further 3 per cent of the authors, and the remaining eighteen developing countries contributed only a further 3 per cent of the authors between them.[19] On this evidence, the research communities of the developing nations are still making only a very slender contribution.

On the other hand, some progress is being made in establishing techniques for arriving at research priorities which are consistent with national development goals. During the 1970s Unesco developed a technique based on Delphi forecasting[20] which enables a nation to determine its priority research areas; representatives of the national scientific community and development programme officials take part in a series of increasingly refined brainstorming sessions from which emerges eventually a list of development goals and the research priorities which should accompany them. Since 1977, the technique has been used in Ghana, Morocco, Jordan, Costa Rica, Argentina, Portugal, and Australia.

Agenda for action

Since the mid-1960s, when the need for national capabilities in science and technology became apparent, there has been considerable progress. Most developing countries have expanded their tertiary education very considerably; most have set up a series of research institutes, both within and outside the university; most have an organizational structure for arriving at, and implementing, science and technology policies; most have established the nucleus of a national scientific community with increased, though still inadequate funding; and most have taken steps to orient their scientific activities towards their national development goals.

In spite of this progress, the results have left much to be desired. At the risk of over-simplifying, and while acknowledging that every solution must be unique to the individual country concerned, most developing countries appear to share a number of major problems at the beginning of the 1980s:

- scientists are underpaid, overworked, and have low social status

- the scientific community remains isolated from international science
- bureaucratic procedures are used to manage research
- an appropriate balance between fundamental and applied research has yet to be struck
- science has not yet been sufficiently popularized to attract widespread interest and support from the people
- nearly all research is government-funded, and private industry has invested little or nothing in research and development
- the link between research, the economy, and society has yet to be established
- the link between research and the unique development goals of each country has also yet to be established
- importation of foreign technology is still seen as quicker and easier than using national science capabilities to adapt and adopt existing local technologies to the same end
- the scientific community plays too small a role in selecting those foreign technologies which it is clearly essential to import.

Future actions to improve the situation are bound to meet one particularly awkward set of problems: many of the issues outlined above tend to reinforce other issues. For example, the more a country tries to link its research with its development goals, the more likely is its research community to remain isolated from the mainstream activities of the international scientific community (though technical co-operation between developing countries (TCDC) could help here).

However, there is one major factor still preventing the developing countries from consolidating their gains, and developing their national science and technology a great deal further. That factor is the degree to which these countries still find themselves in a position of dependency on the developed nations, and the firms operating from their shores. For all the current talk about codes of technology transfer, it remains a fact that about 200 multinational enterprises in the world effectively control nearly all the flow of technology between countries.

11 The transnational enterprise

Such exclusive companies, therefore, are a nuisance in every respect; always more or less inconvenient to the countries in which they are established, and destructive to those which have the misfortune to fall under their government.

Adam Smith, 1723–90

While Adam Smith castigated the monopolies of his time in his famous book *The Wealth of Nations*, the contemporary world has chosen the transnational corporation (TNC) as its equivalent scapegoat. The reason why TNCs are the subject of a whole chapter in this book is simple enough: the TNCs play by far the most important role in the business of transferring technology across national frontiers. Because of this their overall importance in the science, technology, and development equation has become paramount: something like 90 per cent of all technology transfer in the world is accomplished by the TNCs, at a cost to the developing countries of at least $50,000m. a year.

What, exactly, is a TNC? Examples are easier than definitions. Charles H. Taquey provides both when he writes that a TNC is 'an information and decision system that directs the common strategy of business establishments operating under several jurisdictions; its objective is precise and concrete: it is to realize a profit by producing and selling goods and services, computers perhaps or hamburgers, or leisure, under such names as IBM or VW, McDonald or Club Méditerranée'.[1]

It is hard to exaggerate either the size or the power of the TNC in the contemporary world. In all, there are more than 10,000 TNCs, with more than 50,000 foreign affiliates. However, less than 200 of them control more than one-half of international direct investment. The 300 largest TNCs in the United States, for example, have 5,200 foreign subsidiaries. Together, they are responsible for no less than 28 per cent of world exports, including 47 per cent of the exports of primary products and 20 per cent of manufactured products. Many TNCs are themselves economically far more powerful than individual developing countries.

Many of them also conduct businesses which are research-intensive. In fact, some of the larger transnationals spend more on research and development than most developing countries and even some developed ones, as Table I shows.

Unfortunately, little of this research is of direct benefit to developing countries trying to build up their own research programmes. Transnationals tend to conduct their research and development in their home countries. What little they actually carry out in developing countries is usually done by staff from developed countries.

It must be stressed, of course, that the TNC is not just a problem in relation to developing countries. The power of a large TNC, and its ability to operate outside the national legislation of the country where it first started business, combine to produce problems in the industrialized nations as well.

Table I: R and D expenditures by selected countries and corporations (1975, $m.)

Federal Republic of Germany	8,847
Italy	1,656
Sweden	1,216
General Motors	1,114
International Business Machines	946
Belgium	764
Ford Motor Company	748
American Telephone and Telegraph	619
India	420
Spain	262
International Telephone and Telegraph	219
Republic of Korea	127

Source: Norman, Colin, *The God that Limps*, p. 92. New York and London, W. W. Norton, 1981.

European Common Market countries, for example, have complained bitterly about the behaviour of IBM in Europe; counter-complaints have come from the United States about the level of steel exports to the United States from both Japan and Europe; official US policy towards the Soviet Union has been undermined by the insistence of TNCs in Europe and Japan on their own freedom to work on the oil pipeline linking Siberia to Europe; and many European countries as well as the United States have protested vigorously over the degree to which Japanese industry has swamped the market, particularly in fields such as cars and electronics.

And these, after all, are only just a few recent examples of conflicts in the developed world between nations and TNCs. The list could be greatly extended. But whether the way in which a TNC manages to escape the legislation of what was once its host country is in the end a good thing or a bad one is a much more difficult question. Fortunately, it does not have to be answered here; the point is simply that if TNCs are in a position to create this amount of difficulty in the industrialized world, it is hardly surprising that their role in developing nations is far more controversial.

The technological monopoly exerted by the TNCs over developing countries is apparent from the patent distribution. Of some 3.5 million patents in existence in 1972, only 6 per cent were held in developing countries. And five out of these six percentage points were patents held by foreigners. As a result about 1 per cent of the world total of patents was held by nationals in developing countries. It is doubtful whether this figure has risen during the past decade.

The role of the TNC is thus crucial to development. But whether it is crucially good or crucially bad is a matter of opinion—and opinions tend usually to fall into one of two opposed camps: the first holds that TNCs are the key to the future of Third World development and the second, more popular within the United Nations system, that TNCs are the root of all evil.

The fundamental issue can be expressed simply enough:

Those who worry about the poverty of the world and feel that it is the duty of those who are relatively well off to help those who are relatively badly off, have to recognize that the problem cannot be left to private investment. Private investment leaves out the poorest and most needy countries, just as in a national framework it finds no employment for those who are stupid or otherwise economically unproductive.[2]

The critical view, of course, also charges that TNCs, in their zest for maximizing profits and minimizing risks, are in effect holding a pistol to the heads of developing countries. Relationships between TNCs and many developing countries can never be on an equal footing. Very often the TNC holds the strings of power, and it knows it. With their poorly developed scientific and technological communities, the developing countries are in no position to negotiate in a knowledgeable way with a TNC about either the kind of technology the TNC proposes to operate or the long-term effects the technology may bring. The TNCs, because they own the technology, can dictate their own terms to poorer countries which need the technology for development.

TNCs are also charged with other, more environmental crimes. For example, they regularly export chemicals and manufacturing plant which have been banned in their country of origin. They also tend to use developing countries as 'pollution havens' for processes which would be banned elsewhere in the world. As A. J. Dolman reports:

The Third World is not only locked into a system in which it must import technology with inbuilt structures that encourage environmental degradation, it is also increasingly becoming the dumping ground for the dirty production processes that the rich no longer want. These are trends that do not augur well for the future state of the physical environment in the Third World.[3]

For example, the Caribbean has become an important centre for oil refining. The oil comes from Arab states and from Venezuela and is refined in the Caribbean because it has a high sulphur content. The levels of air pollution produced during the refining would not be permitted in the countries where the oil is to be used, such as the United States. Yet the oil produced by Trinidad and Tobago, which is of lower sulphur content, is exported crude to the United States where the TNCs refine it at no pollution cost. Similar stories abound in other fields.

However, the issue of whether or not TNCs should be curbed in this kind of behaviour is not as easy as might at first appear. Those who have suggested international legislation to prevent it have been branded as 'environmental imperialists' by some members of developing countries, who claim they should be free to choose what they wish to buy and what they wish to reject.[4] As Sri Lanka's Ambassador to the UN, H. S. Amerasinghe, has put it, developing countries would be prepared to accept '100 per cent of the developed countries' National Polluting Product, if by so doing they could diversify their economies through industrialization'.[5]

It is for these reasons that the United Nations system has, surprisingly to many people, become heavily involved in the TNC issue. There is now a UN Centre on Transnational Corporations (the CTC) in New York and a UN Commission for Transnational Corporations which has established a working group to draft a Code of Conduct for TNCs. The Centre publishes a periodical on TNCs called *The CTC Reporter*.

A Code of Conduct for Technology Transfer is also in the process of preparation by the UN Conference on Trade and Development (UNCTAD). The main thrust of this document is to try to break the monopoly over technology which is claimed to be held by the TNCs. Among other things, UNCTAD charges that when TNCs export a technology to a developing country they often:

- restrict the fields it can be used in, or reserve alternative uses for themselves or another party other than the developing country
- ban the acquisition of competing technologies as a condition of sale or lease
- place limits on the size of production facilities, the volume of production, or the use of technology in other plants
- sell or license the technology exclusively to one party in a country, thus requiring other producers in the country to pay again for a technology already at work

- sell several technologies in one package, thus forcing developing countries to buy unwanted techniques along with the one they do want
- insist on tied purchases, which means that the developing country is forced to buy raw materials, spare parts, or intermediate products from the TNC at inflated prices. (A 1975 UNCTAD survey found that imported pharmaceutical products were overpriced by 30–500 per cent in Chile, 20–300 per cent in Peru, 40–1,600 per cent in Mexico, and 100–800 per cent in Spain. In Colombia one product was being sold for 55 times the price at which it could be obtained from eight European laboratories)
- insist on excessive quality control as a means of limiting production.

During the 1970s, as part of the thrust towards the New International Economic Order, several different approaches were developed to the problem of placing the developing countries in a better position with regard to the acquisition of technology. While two UN bodies were working on the two Codes of Conduct on TNCs and Technology Transfer, a third—the World Intellectual Property Organization (WIPO, a UN agency with headquarters in Geneva)—was attempting the seventh revision of the Paris Convention (governing protection of industrial property) since it was first signed in 1883. The basis of the revision was to give preferential treatment to developing countries, introduce concessions over access to patent information, and eliminate some of the restrictive practices of the TNCs.

In Latin America six Andean countries signed the Andean Pact in the 1970s which was designed to unite the economic power of the countries concerned. TNCs were thereby required to occupy a minority position in the ownership of companies they set up in those countries, and restrictions were imposed on their degree of dependency on their parent companies. The Pact appears, however, to have had little success. Chile subsequently withdrew from it, Colombia was indifferent to it, and Peru never enforced it.[6]

Unesco also had a study programme related to the TNCs. An

expert meeting held by Unesco in April 1978 concluded that 'Aside from a few positive repercussions, [TNC] activities have negative effects on the social and cultural values and policies, the national identity and the life-style of the host countries . . . The activity of TNCs is one of the forces causing the lag observed between governmental development goals and actual results. External forces conflict or interfere with a process of endogenous development.'[7]

By the beginning of the 1980s it was clear that none of these actions went nearly far enough. Valuable though they may well be, they are aimed only at providing the developing countries with fairer access to existing technologies. They concern the method of technology transfer more than its content. Ann Mattis put her finger on the problem in 1979 when she wrote:

In assessing technology to be applied to development, Third World countries should develop consistency. The 'schizophrenia' often demonstrated in the international arena consists in denouncing with the right hand technological dependencies while with the left concentrating on negotiating for inappropriate technologies free of charge or at cheaper rates. The most appropriate should always be the rule. If a particular technology is bad, that is inappropriate to the need, then it has to be rejected even if it is seemingly made available free of charge. 'Seemingly', because there will always be hidden costs—costs resulting from engaging local resources and local talent for useless production.[8]

Nor do current actions regarding technology transfer look forward to the time when developing countries will be in a position to generate their own technology rather than buy or lease it. During the next two decades the content of technology transfer and the building up of technological resources in developing countries are likely to become the prime focus of attention.

According to Henri Hogbe-Nlend, President of the African Association for the Advancement of Science and Technology, even the idea of technology transfer can do harm:

Use of the term 'transfer of technology' can entail serious misconceptions. It arose when an attempt was made to sanction the idea that all scientific and technological knowledge was generated by certain peoples in certain areas of the world, and that the role of the others was merely to learn it from them. As we all know, this does not correspond to the historical evolution of scientific and technological truth . . . The concept of technology transfer inhibits the scientific and technological creativity of our peoples and aims at placing them in a situation in which they can only mimic, copy and ape, thus virtually condemning them to a position of absolute and permanent dependence.[9]

It is also important to stress that what is popularly called technology transfer is badly named. Very often it is not actually the technology which is transferred so much as the facilities for doing something. As I. H. Abdel Rahman has pointed out, 'You cannot transfer technology. Technology is not transferable. It is acquired . . . What we are seeing now in many developing countries is not technology transfer at all but simply the buying of equipment, of turnkey factories. If someone gives you a motor car you may learn to drive it in five minutes but you do not thereby acquire the technology of making motor cars. You acquire only the technology of using them.'[10]

The problem might not be so serious if the content of technology transfer were more appropriate to the needs of the recipient countries. By and large it is not. For example, a study[11] made during the 1970s by the Economic Commission for Latin America (ECLA) concluded that the technologies imported into Latin America:

- were too automated and capital-intensive, when one of the region's major problems was unemployment
- required highly skilled labour which was not available
- relied heavily on synthetic materials when the region had an abundant supply of natural raw materials
- were too expensive
- perpetuated dependence on Northern suppliers.

While these are the nuts and bolts of the problem, the issue is more profound, reaching down to the fundamental assumptions inherent in the Western life-style. A. J. Dolman, writing of the relationship between the Third World environment and the world economic system, puts it thus:

Every technology carries its own code—economic, social, cultural and environmental. Western technology embodies the code of Western cosmology. The economic code requires that technologies be capital-intensive, research-intensive, organization-intensive and labour-extensive. The social code is one which creates a centre and a periphery and thus perpetuates a structure of inequality. The cultural code presents the Western world as being entrusted with the historic mission of moulding the world according to its own image. The environmental code built into the technology presupposes a relationship between man and nature that is typically vertical; one in which man is given rule over nature, to be to nature what God is to man.[12]

This kind of technology, of course, became dubbed as 'inappropriate' for developing countries during the course of the 1970s. The idea that there existed alternative technologies which were labour-intensive, cheap, environmentally sound, small in scale, easy to understand and operate, did not disrupt local culture, and used only renewable resources had earlier been expounded by many writers, notably the British economist E. F. Schumacher.[13] The debate as to whether this was just another neo-colonial trick to delay development on the part of the industrial nations, or whether it represented a novel and effective solution, raged fiercely throughout the 1970s. The idea was espoused by the prestigious Brandt Commission:

The question of appropriate technology is relevant to both rich and poor countries. Industrialized nations too need more appropriate technologies which conserve energy and exhaustible resources, which avoid rapid job displacement and which do not damage the ecology. It is quite possible that rising energy costs, afflicting both North and South, will eventually compel corporations in the North to concentrate more on new kinds of techniques which may be appropriate to many parts of both North and South. It is a question of enabling the inventiveness and enterprise of scientists and engineers everywhere to give the fullest possible benefit to mankind. To achieve this will require developing countries to be more adaptable, and to assimilate appropriate technology effectively. It will require from the industrial world a greater awareness of the needs of the rest of the world, and greater incentives to extend and adjust their own inventiveness.[14]

Although there was considerable activity in setting up appropriate-technology centres in developing countries during the

1970s, the concept met with continued resistance from much of the Third World. By the time of the UN Conference on Science and Technology for Development in 1979, Anil Agarwal was able to report:

One point that did emerge very strongly during the Vienna conference is that Third World governments are in no mood to accept the idea that small-scale, simple rural-oriented technologies—often diplomatically described as 'appropriate technology'—are best for them. Many in the Third World now see advocacy of appropriate technology as a subtle attempt by the Western countries to undermine the industrial muscle that the Third World is beginning to acquire. The Third World delegations at UNCSTD made it clear that what the developing world really needs is the creation of home-grown capabilities in science and technology to enable governments to make proper decisions about which technology to choose.[15]

There is every chance, however, that the appropriate-technology issue is about to be replaced by what seems likely to be called 'human needs technology'. As development programmes have swung progressively towards helping poverty-stricken rural populations, it has become apparent that the old idea of technology transfer is not only inappropriate but massively so. In many areas technologies to help the truly needy simply do not exist.

If they do not exist, they will have to be invented. And in an era in which most new technology is thought to arise as a result of scientific research, it is obvious that research priorities should be changed. This was one of the factors which led Unesco's Advisory Panel on Science, Technology, and Society to recommend, at its second meeting in late 1981, that 'Science transfer to developing countries should always precede technology transfer, if it is to have lasting benefits and contribute to economic independence.'[16]

One of the legacies of the colonial period is that in many developing countries the research system now in existence has evolved from an older one initially set up under colonial rule. Almost without exception, this means that the research direction is related to industrial applications, usually in the area of cash crops of one kind or another. This is hardly appropriate to the new development era in which the goals of self-reliance, endogenous development, and help for the rural poor are paramount. So what should be the new goals for research in developing countries?

Figures published in the late 1970s indicated clearly where the needs lay. Of a world population of some 4,000 million, there were approximately:

- undernourished   570 million
- adult illiterate   800 million
- children not enrolled in school   250 million
- with no access to medical care   1,500 million
- with less than $90 income per year   1,300 million
- with life expectancy below 60 years   1,700 million
- with inadequate housing   1,030 million

Since 1975 the Unesco programme on 'Research and Human Needs' has been trying to identify where the research priorities lie. A series of indicators has been developed to help each developing country determine its priority needs, and a list of corresponding research projects has been developed. There are, of course, many problems to this approach.

First, research is currently conducted by discipline rather than by problem area. Hence there are difficulties associated with the formation of multidisciplinary teams to tackle individual problems. Partly for this reason, the Unesco Advisory Panel on Science, Technology, and Society has recommended that all Unesco's science and technology programmes be problem-oriented rather than disciple-oriented.[17]

Secondly, the problems themselves are often not amenable to discrete solution. In other words, they are mostly dependent on one another—for example, solutions to problems of biomass energy production may depend on first solving the problems of expensive fertilizer. According to the first report of the Unesco programme on 'Research and Human Needs', prepared in conjunction with the International Federation of Institutes of Advanced Study, 'contemporary problems exist as an untidy tangle of intertwining difficulties which is now often referred to as the "world *problématique*" '.[18]

Thirdly, the time-scale between research and application can be very long. And, fourthly, it is essential that needs are defined by the developed country itself, not by a technical adviser from the industrialized countries. The aim of the Unesco programme is to provide developing countries with the kind of information required to establish what their needs actually are, and hence to determine the shape of appropriate research programmes. It is hoped that it may also stimulate the research community in developed countries to reorientate their work towards issues more relevant to the needs of developing countries.

But how can this be achieved? It is here that the international community comes into its own. The success or failure of the attempt to stimulate the endogenous development of science and technology in developing countries depends ultimately on international organizations, both those within and those without the United Nations system. However local the solutions that may be developed, they stem from a global understanding of the real issues involved. The international dimension is the key to the future.

12 The international dimension

Any man's death diminishes *me*, because I am involved in *Mankind*; And therefore never send to know for whom the *bell* tolls; It tolls for *thee*.

John Donne, 1571–1631

The world of the 1980s is a confusing place. World expenditure on armaments, and on the science and technology which produce them, is higher than ever before. Despite a world recession, the developed countries are becoming increasingly involved with a number of glamorous new technologies, particularly in the information field: within a space of perhaps only a decade the lives of their populations are likely to be altered dramatically by the advent of new types of information system depending ultimately on the microchip. Yet expenditure on research, which has made all this possible, is declining in real terms.

Enormous disparities still exist within the science and technology system. In the developing countries less than one person in 100,000 is a scientist. In some countries—such as Argentina, Brazil, and India—subsantial scientific communities exist, even if they lack bite and are not yet accorded sufficient recognition by their governments. But in others, indeed in most developing countries, what passes for a scientific community is still pitiful—a small group of dedicated and talented people constantly battling for financial support, without formal organizations, with scant access to the outside world of science, and often without even the means to travel to important international conferences.

It is difficult to be optimistic about the future. To be sure, the idea of the endogenous development of national science and technology has now been firmly implanted, thanks partly to the UN Conference on Science and Technology for Development in 1979 and partly to the insistence on the importance of national science and technology in the New International Economic Order.

Yet little enough seems to get done. In one sense, of course, the problem is one which the developing countries themselves must wrestle with. As Abdus Salam, the Pakistani physicist, has pointed out, 'The development of science and technology in developing countries is ultimately and squarely the responsibility of the countries themselves.'[1] But he insists that they must recognize that:

- the development of science is not cheap and requires at least 1 per cent of GNP
- a scientific community must attain a critical size before it produces results—typically, every one person in a thousand needs to be a scientist, as in the developed countries
- there can be no leap-frogging in applied science if basic sciences are weak
- science is international, and science in any region which is isolated from outside contact will die
- science must be administered by scientists and not by bureaucrats or economists in planning commissions.

Salam has a message for the developing countries:

Your men of science are a precious asset. Prize them, give them opportunities, responsibilities for the scientific and technological development of their own countries. The goal must remain to increase their numbers tenfold, to increase the $2000 million spent on science and technology to $20,000 million. Science is not cheap; but expenditure on it will pay tenfold.[2]

One important lesson, however, may have been learned during the 1970s. As our perceptions of our global environment have changed, so too have the links which bind all countries and all peoples together become more visible. It is, of course, morally indefensible that disparity and dependency should still exist in the world on such a massive scale. But moral questions apart, it is now also apparent that inequities of this order are not only bad for the underprivileged: they are bad for everyone. Society cannot exist permanently with this kind of imbalance. Stability and growth will come only with equity.

Thus while the main responsibility for fostering endogenous science and technology must rest with the developing countries themselves, everything possible must be done by other actors on the international stage to help them. In the world science scene, there are hundreds, perhaps thousands, of actors involved—from national science communities and regional associations through to international non-governmental organizations such as Pugwash and the international agencies themselves.

If any one agency has specific responsibility for the international development of science it is Unesco, the only agency within the United Nations system charged specifically with fostering international scientific co-operation. Yet it must be stressed that the funds available to an organization such as Unesco for such a massive responsibility are extremely limited. The science sector in Unesco, which is responsible for many different programmes, including one on international scientific co-operation, commands an annual budget little larger than that of a single faculty in a large university in an industrialized country (less than $25m. a year).

There are nevertheless techniques available to Unesco—some old, some new—which are capable of catalytic action: small sums judiciously used can ignite a chain reaction of co-operation with ultimate benefits, particularly to developing countries, which are out of all proportion to the original cost. Identifying these techniques, and making them work efficiently, is no easy matter. Yet it can be done. It is, as Salam has eloquently put it, a question of identifying those places in the world—or of creating them— where one may 'light a candle from a candle'.[3]

Consider first the case of the International Centre for Theoretical Physics (ICTP) in Trieste. The brain-child of Abdus Salam, the ICTP was conceived as a place where promising young physicists, at least half of them from developing countries, could work for six weeks or more a year while on vacation from their own universities. In the process, they would meet with the best minds in their subjects in the world. As the centre was theoretical, no equipment, other than blackboard and chalk, was required.

The centre was created in 1964 by the International Atomic Energy Agency (IAEA), with help from Unesco and generous assistance from the Italian government and the city and province of Trieste. In 1970 Unesco became a full partner in the venture. Every year hundreds of physicists have benefited from the centre's 'associateships' by attending its symposia and workshops. The centre has become a major success and fulfilled a major need.

Interestingly, as the years have gone by, the subjects of interest to the centre have changed. The emphasis has altered from a purely theoretical approach to a concentration on more technological, practical issues, albeit ones closely related to the growing points of advanced physics. This is a key issue, and one which we shall return to later in this chapter.

Yet, as the centre's success becomes more assured, so does its future become more uncertain. The centre's budget was provided at the beginning of the 1980s by a 50 per cent contribution from the Italian government, the rest being shared equally between the IAEA and Unesco. But, as already mentioned, Unesco's job is to help create such schemes, not provide them with long-term support. The sad fact remains that if Unesco withdraws its contribution, which could now be better spent in catalysing the creation of other, similar centres, the ICTP could well die for lack of support. So far, no international funding body has been identified which is willing to take over the responsibility on a permanent basis. Until one is found, the ICTP must spend precious time fighting for its financial survival.

The ICTP is the only body of its kind available to physicists from something like 100 developing countries. Its budget is of the order of $1.5m. a year. Yet the expenditure in Europe on joint projects in physics is well over $500m. a year. As Salam ruefully points out, 'Somehow, somewhere, a break must come.'[4]

Salam has one idea of where the break might come from. In 1979 he made an appeal to his brothers in Islamic countries:

To some of you Allah has given a bounty—an income of the order of

$60,000 million. On the international norms these countries should be spending $1000 million annually on science and technology. It is their forbears who were the torchbearers of international scientific research in the 8th, 9th, 10th and 11th centuries. It was these forbears who funded Bait-ul-Hikmas—Advanced Institutes of Sciences—where concourses of scholars from Arabia, Iran, India, Turkey and the Byzantium congregated. Be generous once again. Spend the billion dollars on international science, even if others do not. Create a Talent Fund— available to all Islamic, Arab and developing countries—so that no potential high level talented scientist is wasted.[5]

So far, the appealing prototype of the International Centre for Theoretical Physics in Trieste has not proved to be a candle from which others can be lit. Financiers, Islamic or otherwise, have not appeared to fund a score of similar centres for other fields. But fortunately the ICTP model is only one of the ways in which new international initiatives can be launched.

The most common alternative, of course, is the international research programme, the oldest form of international collaboration known to science. Science is international partly because its subject-matter is. And though scientific institutions may be born first and foremost with national objectives in mind, they rarely remain as such for long. In particular, any research to do with the planet itself quickly demands global attention.

The first attempt at international collaboration in science took place more than a century ago. However, it was not until the highly successful International Geophysical Year (IGY) of 1957/8 that the idea of large-scale international co-operation devoted to specific themes became really popular. Organized by the International Council of Scientific Unions (ICSU), the IGY involved the co-operation, over an eighteen-month period, of scientists from sixty-five different countries, twenty international organizations, and more than one hundred national centres. One of the major discoveries of that venture was the Van Allen belts which surround the Earth, the existence of which has since explained a great deal about the behaviour of the earth's magnetosphere.

Since then, intergovernmental research themes have become almost common. Unesco's own Arid Zone programme was continued until 1962. Its success, not least in the developing countries, gave rise to many others—the International Biological Programme, Man and the Biosphere, the International Hydrological Programme, the Intergovernmental Oceanographic Commission, the International Geological Correlation Programme, the Long-Term and Expanded Programme of Oceanic Exploration and Research (LEPOR), and UNISIST (the United Nations programme for information on science and technology).

In programmes of this kind, a number of countries, typically about thirty, agree to co-operate in research of a specific type. They provide their own scientists and research centres to whatever extent they are able. They do so under the guidance of a governing scientific body which has a permanent secretariat. This is virtually the only additional cost of such schemes, though individual countries may well increase funding of science in the area concerned to improve their own contribution. In major programmes additional support is also channelled through the international agencies such as Unesco.

The common features of these programmes are that:

- they deal with problems which are global by nature
- they interest both developed and developing countries as they concern improved understanding of regional or global phenomena, and their management
- though they are internationally co-ordinated, they are executed by national centres, usually under the guidance of national committees
- they are not academic but problem-oriented
- they include not only research but also training which is essential if all relevant countries are to make an active contribution.

Two points need spelling out in more detail. The first is that throughout much of this book the emphasis has thus far been on research. Research is, in fact, only one of two essential ingredients in scientific activity. The other is education and training. There can be no science without scientists, and no scientists without appropriate education at all levels from the primary school to the university. No country can build up its scientific and technological community without the necessary educational apparatus, and it is for this reason that Unesco devotes so much attention to scientific and technological education.

But the educational effort must be well balanced. Over the past decade or so many developing countries have spent enormous sums on tertiary education. Huge university complexes have grown up, and the output of qualified people has grown considerably. Unfortunately, this has not always produced the desired effect of strengthening the scientific capabilities of those countries. One reason is that sometimes prestigious university complexes are financially favoured over primary and secondary educational needs. The inevitable result is that to attract sufficient students, the universities have had to lower their standards. Because the capabilities of a scientific community are better measured in terms of excellence than sheer size, the result has left much to be desired.

The second point which needs emphasis is that intergovernmental research programmes concern problems rather than academic subjects. This is important, for problem-solving is more attractive to government finance than is the support of subject areas by discipline. Indeed, one of the problems of science as a whole has always been to attract funding for academic research; applied studies are more easily funded.

If the future borderlines between academic and applied science, or between science and technology, become more blurred, one aspect of research funding may at least be eased, and the old debate about the value of pure research will finally be made redundant. Perhaps this is one reason for the current popularity of the earth sciences where, as we saw in Part II (Chapter 6, 'Natural systems'), it is now virtually impossible to classify ongoing science as either applied or pure. Indeed, it is increasingly difficult to distinguish the activity which provides scientific information from activity which utilizes that information in order to improve the management of the Earth's natural systems. As Unesco's Advisory Panel of Science, Technology, and Society put it: 'The earth sciences are merging with land-use planning and conservation of the natural environment.'[6]

Most of the international research programmes are co-ordinated by a specially created body, such as the Intergovernmental Oceanographic Commission in the case of LEPOR. However, few of them would have been possible without the existence of ICSU, the International Council of Scientific Unions, and all the scientific unions for nearly every scientific subject which are the members of ICSU. These unions are the best sources of scientific information and expertise in the world.

Until recently, however, they have concentrated most of their efforts in the developed countries. Close relationships exist between Unesco and other international organizations and the scientific unions (indeed, Unesco helps finance ICSU). In the future, however, it is clear that the unions should do all in their power to foster the development of science and technology in developing countries. This may be one of the key ways of breaking down the sense of isolation felt in the embryonic scientific communities of the Third World. Similarly, it may be possible to extend the ways in which seed money is being used to help establish national and regional associations of scientists in developing countries. The sense of belonging to a community is one of the vital ingredients still missing for many scientists in developing countries.

Yet another way of providing this missing link is to form networks of co-operation between laboratories working on common problems in developing countries. These networks can either have a permanent focal point in the form of a selected institute in one country, or a rotating focal point which changes from year to year. Either way, networks offer excellent and cheap means of fostering international co-operation. Recently, successful networks in such fields as microbiology and the chemistry of natural products have been set up. Ideally, these networks should concentrate on co-operation in both research and training. They should also act as clearing-houses for information. One of their great advantages is that because they offer individual scientists contact with other institutes working in the same area of science on a relatively informal basis, the danger of permanent emigration via the brain drain to a developed country is much reduced.

Networks exist in a number of different regions and in a number of different subjects—but there is a need for more, particularly in such areas as rural-development technology and low-cost housing. Where it seems difficult to set up a regional network at first, it may often be possible to set up national networks, linking the institutions working on common problems within a single country. Once this has been done, it is often easier to link the national networks together to form a regional or world-wide network.

A good example of a regional network is the South-east Asian Network for the Chemistry of Natural Products, set up in 1976 and financed by an annual grant from Unesco and the Japanese government. It has ten member nations, with a headquarters in Bangkok. It offers qualified scientists in those countries a free return air fare to an institution in the region, free accommodation, a textbook allowance of $50, and $500 a month for general expenses. It governs its own affairs through a Co-ordinating Board consisting of one chemist from each member nation and publishes a newsletter three times a year called *Chemistry in Asia*. In many ways it offers an excellent model for networks in other regions and in other subject areas.

There is a similar south-east Asian network for the micro-biology of natural products. Both networks are founded primarily on the principles of mutual co-operation and self-help. And as well as offering the chance of travel to individual scientists, they hold regular training courses and workshops. More than fifty have already been held, involving some 1,000 trainees and 150 lecturers recruited from all over the world.

Unesco's Advisory Panel on Science, Technology, and Society has made a number of key recommendations about the activities of networks.[7] Among other things, the panel suggests that networks should:

- organize training workshops and travelling seminars
- create expert groups to help on development issues

- stimulate basic and applied research in parallel, with an emphasis on transforming result into productive work
- take special care to bridge language and development differences in their training programmes
- exchange reprints and non-commercial data bases.

Among the ways in which Unesco could specifically help is by a request to member nations to use their data bases free of charge to serve researchers in developing countries.

These, then, are some of the ways in which an international organization such as Unesco can help foster international collaboration in science, particularly among the developing countries. There are, of course, many others—such as the awarding of prizes to individual scientists of exceptional merit, the financing of individual research or training schemes in particular areas, and implanting seed money in institutes which appear to have a chance of becoming centres of excellence with a little help.

Many other suggestions have been made. One particularly attractive one is that an institute of advanced systems analysis for North–South problems be created. The International Institute of Applied Systems Analysis in Vienna has already made a number of major contributions in this area, but mainly either on global problems or on problems related specifically to the needs of developed countries. There is certainly a real need for a parallel body to assess the problems of the developing world, and its relationship to the developed countries.

There is also a need for action in the general area of the science–technology relationship. So far most attention has been given to international collaboration in the area of science itself—partly for the obvious reason that the exchange of scientific information is, at least in principle, free and unhampered, while technological information is bounded by both commercial and military secrecy and, thus, by strong elements of competition.

In the world of the future, there are certain to be changes. One of the most important may be the gradual merging of pure and applied science to produce an activity which is simultaneously both science and technology. Certainly the old idea of a long time gap between research and industrial application, while perhaps still relevant in some areas, is obsolete in many of the contemporary areas of rapid advance.

In such fields as information technology, biotechnology, and new energy sources, new methods of dealing with new issues will have to be (and are being) found. The existing patterns of scientific co-operation will not suffice to deal with the explosive growth of fields of scholarship which defy categorization as either simple science or simple technology. A start will have to be made very soon on linking the world's technological communities together, despite the problems created by restrictive practices connected with the free flow of technological information; the existence of the United Nations Industrial Development Organization, UNCTAD, the World Intellectual Property Organization, and even GATT may be thrusts in the right direction.

In all this, the wisdom of historical insight is sure to play a role. Embedded as we are in the ideas of the second half of the twentieth century, it is difficult to appreciate that ideas which now appear to be part of some permanent truth may not seem to be so in twenty years' time and have not, in fact, been true for very long. Among these suppositions are the ideas that the natural centre of gravity of science and technology lies in Europe and the United States; that the world is forever doomed to be divided into two halves, the 'haves' and the 'have-nots', with one half in a permanent state of dependency on the other; and that development is a never-ending problem.

In considering such issues, one of the aphorisms of the 1960s appears highly appropriate: 'Constant change is here to stay.' In fact, constant change has been around for a long time. For five centuries or more, Greece was the centre of scientific excellence. During the middle of the first millennium Chinese scientists held sway. During the 350 years which followed, the scientists of Islam—Arabs, Turks, Afghans, and Persians—made nearly all the running. Between 1100 and 1350 the first Western men of science began to share the honour with their eastern counterparts. But even in 1720 Maharaja Jai Singh of Jaipur was correcting Western tables of eclipses of the sun and moon. However, as one contemporary chronicler wrote, 'With him on the funeral pyre, expired also all Science in the East.'[8]

Since then Western science and technology have dominated the world. They will not always do so. The job now is to ensure that the fruits of science and technology are evenly dispersed through the world. For that to happen, their seeds must be carefully sown in the fertile ground of the developing countries—not only by those able and willing to support the developing countries but by the developing nations themselves.

Notes

Introduction

1. Batisse, Michel, 'The Relevance of MAB'. In *Environmental Conservation* 7, 3, Autumn 1980, 179.
2. Unesco, *Second Medium-Term Plan (1984–1989)*. Paris, Unesco, 1983, 4 XC/4 Approved.
3. Auger, Pierre, *Current Trends in Scientific Research*. Paris, Unesco, 1961.
4. Buzzati-Traverso, Adriano, *The Scientific Enterprise, Today and Tomorrow*. Paris, Unesco, 1977.
5. In this context world development is used to mean the evolution of society on a world-wide basis, including social and cultural development, in contrast to 'development' which is usually used to signify the improvement of material conditions in the poorer countries.
6. Nelson, Richard R., and Langlois, Richard N., 'Industrial Innovation Policy: Lessons from American history'. In *Science* 219, 18 Feb. 1983, 814.
7. United Nations Environment Programme, *Major Environmental Trends to be Addressed by UNEP, 1982–1992*. Nairobi, UNEP, 1981.
8. It can be argued that the introduction of the railways in the nineteenth century, an engineering innovation based on the earlier non-scientific invention of the steam-engine, changed society more radically than has any other single innovation. However, the process of change was long and drawn-out; the effects of science-based innovation are often, though not always, widely felt within a decade.
9. FAO, *Wood for Energy*. Rome, FAO, 1983, Forestry Programme Priorities No. 1.
10. It should not be forgotten, however, that a modern business selling traditional products may well exploit science-based technology in the production line, in marketing, and in financial control.
11. The analogy that used to be drawn between science and the arts is best forgotten. If pure science were to be subsidized at the same rate as are the arts in contemporary society, fundamental research would disappear virtually overnight.

12. Buzzati-Traverso, *op. cit.*, p. 28.

13. See, for example, Galtung, J., *The True World: A transnational perspective.* New York, The Free Press, 1980.

14. Zuckerman, S., *Nuclear Illusion and Reality.* London, Collins, 1982.

15. Redirecting science and technology towards the goal of development is made more complex by the fact that even in the developed countries the focus of effort is heavily distorted: 'At present, over 40 per cent of the world's best qualified scientists and engineers work towards objectives stemming from the defence requirements of different countries' (Menon, M. G. K., 'Science and Technology for Development: A turning point'. In Standke, K.-H., *Science, Technology and Society.* New York and Oxford, Pergamon Press, 1981).

16. See, for example: 'Research Trends and Priorities in relation to Human Needs Problems'. Paris, Unesco, 1981, SC-81/WS/115. Richardson, J., 'What the Public Needs to Know about Science and Technology: A report to the Club of Vienna'. In *J. Technical Writing and Communication* 11.4, 1981, 303. Advisory Panel on Science, Technology, and Society, 'Contribution to the Science and Technology Part of Unesco's Medium-Term Plan'. Paris, Unesco, 1981, SC/631/03.

17. See *Technology on Trial.* Paris, OECD, 1979.

Part I. Concepts Chapter 1. Science for knowledge

1. Ziman, John, *Reliable Knowledge: An exploration of the grounds for belief in science.* Cambridge, Cambridge University Press, 1978.

2. Granger, John V., *Technology and International Relations.* San Francisco, Freeman, 1979.

3. Ziman, *op. cit.*, p. 1.

4. Quoted in Morison, Robert S., 'Introduction'. In Holton, Gerald, and Morison, Robert S. (eds.), *Limits of Scientific Inquiry*, p. xi. New York, Norton, 1978.

5. Graham, Loren R., *Between Science and Values.* New York, Columbia University Press, 1981.

6. For example, Abraham Maslow has suggested that a subjective science would be useful in some fields of psychology and psychiatry, J. R. Ravetz that a critical science would help curb the worst excesses of inappropriate technology and, in the United States, the New Alchemy Institute has proposed a science which could be carried out by the people instead of professional scientists. However, these isolated proposals are structurally different from those related to appropriate technology. The latter can be combined to form a

coherent technology which answers all the points made by the critics. Proposals for appropriate science cannot be thus combined, and so far appear capable of answering individual criticisms only on a piecemeal basis.

7. See, for example: Kuhn, T. S., *The Structure of Scientific Revolutions*. Chicago, University of Chicago Press, 1962. Lakatos, Imre, and Musgrave, Alan (eds.), *Criticism and the Growth of Knowledge*. Cambridge, Cambridge University Press, 1970.

8. Roszak, Theodore, *Where the Wasteland Ends: Politics and transcendence in post-industrial society*. Garden City, NY, Doubleday, 1972.

9. Eddington, A. S., *Science and the Unseen World*. New York and London, Macmillan, 1929.

10. Anshen, Ruth Nanda. 'Convergence'. In Cavalieri, Liebe F., *The Double-Edged Helix: Science in the real world*. New York, Columbia University Press, 1981.

11. Sinsheimer, Robert, 'Comments'. *Hastings Center Report* 6, Aug. 1976.

12. Siekevitz, P., 'Scientific Responsibility'. In *Nature*, 227, 1970, 1301.

13. Einstein, Albert, 'Science and Religion'. In *First Conference on Science, Philosophy, and Religion*. New York, 1941.

14. Buzzati-Traverso, Adriano, *The Scientific Enterprise, Today and Tomorrow*. Paris, Unesco, 1977.

15. Crick, Francis, *Of Molecules and Men*. Seattle and London, University of Washington Press, 1966.

16. Ibid., p. 95.

17. Graham, *op. cit.*, p. 329.

18. Danzin, André, 'Enterprise et évolution'. In *Connaissance politique*, No. 1, Feb. 1983.

19. Sinsheimer, Robert L., 'The Presumptions of Science'. In Holton and Morison, op. cit, p. 30.

20. US National Academy of Sciences, *Science and Technology: A five year outlook*. San Francisco and Oxford, W. H. Freeman, 1979.

21. Sinsheimer, op. cit., 1979, pp. 34–5.

22. Quoted in Berg, Paul, 'Research with Recombinant DNA'. *Annals of the New York Academy of Sciences*, Mar. 1977, 278.

23. Quoted in Sardar, Ziauddin, 'Why Islam needs Islamic Science'. *New Scientist*, 95. 1299, 1 Apr. 1982.

24. Granger, op. cit., pp. 37–8.

Chapter 2. Science for technology

1. Dickson, David, Presentation to the NGO Forum on Science and

Technology for Development, New York, Sept. 1978.

2. Food and Agriculture Organization, *The State of Food and Agriculture, 1972*. Rome, FAO, 1972.

3. Langrish, J., 'Does Industry Need Science?'. In *Science Journal*, Dec. 1969, 81–4.

4. US National Academy of Sciences, *Science and Technology: A five year outlook*. San Francisco and Oxford, W. H. Freeman, 1979.

5. Bhagavantam, S., 'Improving Science Education'. In Kinnon, Colette M. (ed.), *Scientists on Development: International scientific and technological co-operation today and tomorrow*. Paris, Unesco, May 1979, SC-79/WS/77(rev).

6. Papa Blanco, F. F., *La Encrucijada Tecnologica*, p. 54. Montevideo, Fundacion de Cultura Universitaria, 1979.

7. King, Alexander (ed.), *The State of the Planet*. Oxford and New York, Pergamon Press, 1980.

8. Quoted in Granger, John V., *Technology and International Relations*, p. 117. San Francisco, W. H. Freeman, 1979.

9. Advisory Panel on Science, Technology, and Society. Paris, Unesco, 1981.

10. Gottstein, Klaus, personal communication to the Assistant Director-General for Science, Unesco (14 Mar. 1983).

11. Holton, Gerald, 'The "Double Service" of Scientists and Engineers'. In Kinnon, op. cit., p. 92.

12. Turchenko, V. N., *The Scientific and Technological Revolution and the Revolution in Education*. Moscow, 1973.

13. Medawar, Peter, 'On the Effecting of All Things Possible'. *New Scientist*, 4 Sept. 1969, 465–7.

14. Rahman, Ibrahim Helmi Abdel, 'Science and Technology: The development dilemma'. *Unesco Courier*, Nov. 1979, 4–9.

15. Cavalieri, Liebe F., *The Double-Edged Helix: Science in the real world*. New York, Columbia University Press, 1981.

16. Polanyi, M., *The Republic of Science: Its political and economic theory*. Chicago, Chicago University Press, 1962.

17. Unesco, *Societal Utilization of Scientific and Technological Research*. Paris, Unesco, 1981, Science Policy Studies and Documents No. 47.

18. Collingridge, David, *The Social Control of Technology*. Milton Keynes, The Open University Press, 1981.

19. Cavalieri, op. cit., p. 83.

Chapter 3. Technology for development

1. 'Reaching Beyond the Rational'. *Time*, 23 Apr. 1973.

2. Denison, Edward F., with Poullier, Jean-Pierre, *Why Growth Rates Differ: Postwar experience in nine Western countries*. Washington DC, The Brookings Institution, 1967.

3. US National Academy of Sciences, *Science and Technology: A five year outlook*. San Francisco and Oxford, W. H. Freeman, 1979.

4. Examples of this class of problem might include the provision of rapid transit for large numbers of people in city centres, the elimination of human disease and the perfection of computer language translation.

5. Unesco, 'Contribution to the Science and Technology Part of Unesco's Medium-Term Plan'. Paris, Unesco, 1981, SC/631/03.

6. Presumably the arms race is also an example of an 'incorrect use of S and T'. However, to include it as an example of a problem which is not amenable to science and technology seems contentious. As many observers have pointed out, when so much of current research and development is devoted to the problems of defence, it is not unreasonable to claim that the arms race *is* contemporary science and technology. Colin Norman writes: 'The feeding of the world's military machine is thus the predominant occupation of the global research and development enterprise'—see Norman, Colin, *The God that Limps: Science and technology in the eighties*, p. 72. New York and London, Norton, 1981.

7. Jackson, Michael W., 'Science and Depoliticization'. In Richardson, J. (ed.), *Integrated Technology Transfer*, pp. 141–9. Mt. Airy, Maryland, Lomond Books, 1979.

8. Jackson, op. cit., p. 147.

9. Unesco, *Estimation of Human and Financial Resources Devoted to R and D at the World and Regional Level*. Paris, Unesco Office of Statistics, 1979.

10. Norman, Colin, *The God that Limps: Science and technology in the eighties*, p. 71. New York and London, Norton, 1981.

11. Norman, op. cit., p. 72.

12. Papa Blanco, F. F., *La Encrucijada Tecnologica*, p. 48. Montevideo, Fundacion de Cultura Universitaria, 1979.

13. Singer, Hans, *Technologies for Basic Needs*. Geneva, International Labour Organization, 1977.

14. United Nations, *World Plan of Action for the Application of Science and Technology to Development*. New York, United Nations, 1971. A popular account is to be found in Clarke, Robin, *The Great Experiment: Science and technology in the Second Development Decade*. New York, United Nations, 1971.

15. Chagula, W. K., 'Statement to UNCSTD'. In Standke, Klaus-Heinrich, and Anandakrishnan, M., *Science, Technology and Society: Needs, challenges and limitations*, pp. 590–4. New York and Oxford, Pergamon Press, 1979.

16. Independent Commission on International Development Issues, *North–South: A programme for survival*, pp. 197–8. London, Pan Books, 1980.

17. Patel, Surendra, J., 'Postface'. In Richardson, op. cit., pp. 151–6.

18. Sagasti, Francisco, R., 'Knowledge is Power'. *Mazingira*, No. 8, 1979, 28–33.

19. OECD, *North–South Technology Transfer: The adjustments ahead*. Paris, OECD, 1981.

20. King, Alexander (ed.), *The State of the Planet*. Oxford and New York, Pergamon Press, 1980.

21. Quoted in Granger, John V., *Technology and International Relations*, p. 105. San Francisco, Freeman, 1979.

22. da Costa, Joao Frank, 'Twelve "Musts" for Development'. *Unesco Courier*, Nov. 1979, 10.

23. Unesco, *Thinking Ahead: Unesco and the challenges of today and tomorrow*. Paris, Unesco, 1977.

Chapter 4. Development for what?

1. Casimir, H. B. G., 'Historical and Cultural Perspectives of Science and Technology in the Development Process'. In Standke, Klaus-Heinrich, and Anandakrishnan, M. (eds.), *Science, Technology and Society: Needs, challenges and limitations*, pp. 28–33. New York and Oxford, Pergamon Press, 1979.

2. UN General Assembly Resolution 3362 (S-VII) of 16 September 1975.

3. The United Nations Conference on Science and Technology for Development, *The Vienna Programme of Action for Science and Technology for Development*. New York, United Nations, 1979.

4. Herrera, Amilcar, O., 'A New Role for Technology'. In *Mazingira*, No. 8, 1979, 35–40.

5. Quoted in ibid. 37.

6. Singer, Hans, *Technologies for Basic Needs*. Geneva, International Labour Organization, 1977.

7. Herrera, op. cit., 38.

8. da Costa, Joao Frank, 'Twelve "Musts" for Development'. In *Unesco Courier*, Nov. 1979, 10.

9. Organization de l'unité Africaine, *Plan d'action de Lagos pour le développement économique de l'Afrique 1980–2000.* Geneva, Institut international d'études sociales, 1981.

10. Urquidi, Victor L., 'Science, Technology and Endogenous Development: some notes on the objectives and the possibilities'. Personal communication, 1982.

11. Ibid., p. 22.

12. Cavalcanti, Geraldo Holanda, 'Economic and Scientific Development'. In Kinnon, Colette M., *Scientists on Development: International scientific and technological co-operation today and tomorrow.* Paris, Unesco, May 1979, SC-79/WS/77(rev).

13. See Sardar, Ziauddin, 'Science for the People of Islam'. In *New Scientist* 93.1290, 28 Jan. 1982, 244.

Part II. Trends

1. See Clarke, Robin, and Palmer, Judi, *The Human Environment: Action or disaster.* Dublin, Tycooly International, 1983.

2. United Nations Environment Programme, 'Common Perceptions of Environmental Issues'. Nairobi, UNEP, 1981.

Chapter 5. Basic sciences

1. US National Academy of Sciences, *Science and Technology: A five year outlook*, p. 63. San Francisco and Oxford, W. H. Freeman, 1979.

2. Davis, Bernard D., 'Frontiers of the Biological Sciences'. In *Science* 209.4452, Centennial Issue, 4 July 1980, 78.

3. Bromley, D. A., 'Physics'. Ibid. 110.

4. Rubin, Vera C., 'Stars, Galaxies, Cosmos: The past decade, the next decade'. Ibid. 64.

Chapter 6. Natural systems

1. See Likens, Gene E. (ed.), *Some Perspectives of the Major Biogeochemical Cycles: Scope 17.* New York, John Wiley, 1981.

2. Howarth, M. K., 'Continental Drift and Plate Tectonics'. In Cocks, L. R. M. (ed.), *The Evolving Earth: Chance, change and challenge*, p. 154. Cambridge, Cambridge University Press, 1981.

3. Bishop. A. C., 'The Development of the Concept of Continental Drift'. Ibid., p. 163.

4. Food and Agriculture Organization, *The State of Food and Agriculture 1977.* Rome, FAO, 1978.

5. Organization for Economic Co-operation and Development, *The State*

of the Environment: An appraisal of economic conditions and trends in OECD countries. Paris, OECD, 1979, ENV/Min (79)1.

6. United Nations, *UN Conference on Desertification: Round-up, plan of action and resolutions.* New York, United Nations, 1978.

7. Das, D. C., 'Soil Conservation Practices and Erosion Control in India—a case study'. In *FAO Soils Bulletin*, No. 33, 1977.

8. United Nations, op. cit.

9. Arnold, J. E. M., and Jules, Jongma, 'Fuelwood and charcoal in developing countries: An economic study'. In *Unasylva*, 29.118, 1977, 2–9.

10. Food and Agriculture Organization, Committee on Agriculture, *Soil and Water Conservation.* Rome, FAO, Nov. 1980.

11. Food and Agriculture Organization, *Tropical Forest Resources Assessment Project 1981.* Rome, FAO, 1981.

12. FAO/UNEP, *Pilot Study on Conservation of Animal Genetic Resources.* Rome, FAO, 1975.

13. Olmo, H. P., 'Grapes'. In Simmonds, N. W. (ed.), *Evolution of Crop Plants.* London and New York, Longman, 1976.

14. Tolba, M., Speech to the Environment Committee of the European Parliament, 25 Nov. 1981.

15. International Union for Conservation of Nature and Natural Resources, United Nations Environment Programme and World Wildlife Fund, *World Conservation Strategy.* Gland, Switzerland, 1981.

16. Ibid.

17. See, for example: Bourlière, F., and Batisse, Michel, 'Ten Years after the Biosphere Conference: From concept to action'. In *Nature and Resources* 14.3, 14. Batisse, Michel, 'The Relevance of MAB'. In *Environmental Conservation* 7.3, Autumn 1980.

18. Some of the environmental catastrophes produced by development projects are collected together in one book which encapsulates a widely held view of the relationship of technology and the environment in the early 1970s—see Farvar, M. Taghi, and Milton, John P., *The Careless Technology: Ecology and international development.* New York, The Natural History Press/Garden City, 1972.

19. Batisse, Michel, 'The Biosphere Reserve: A tool for environmental conservation and management'. In *Environmental Conservation* 9.2, Summer 1982.

20. Udvardy, M. D. F., 'A Classification of the Biogeographical Provinces of the World'. Gland, IUCN, 1975, IUCN Occasional Paper No. 18.

21. Batisse, M., op. cit., 1982, 107.

22. Figures on the world hydrological balance are taken from *The Global 2000 Report to the President*. Washington DC, US Government Printing Office, 1980. Slightly different figures can be found in the other major sources, which include: Baumgartner, A., and Reichel, E., *The World Water Balance*. Vienna, R. Oldenbourg, 1975. Lvovich, M. I., *The World's Water*. Moscow, Mir Publishers, 1973. Unesco, *World Water Balance and Water Resources of the Earth*. Paris, Unesco, 1978.

23. See 'Managing Our Fresh-Water Resources'. In *Impact of Science on Society*, No. 1, 1983.

24. For a history of the Intergovernmental Oceanographic Commission, see Roll, Hans Ulrich, *A Focus for Ocean Research: Intergovernmental Oceanographic Commission—history, functions, achievements*. Paris, Unesco, 1979, Intergovernmental Oceanographic Commission Technical Series No. 20.

25. For a popular account of the International Indian Ocean Expedition, see Behrman, Daniel, *Assault on the Largest Unknown: The International Ocean Expedition 1959–65*. Paris, The Unesco Press, 1981.

26. GESAMP, *The Health of the Oceans*. Geneva, UNEP, 1982, UNEP Regional Seas Reports and Studies No. 16.

27. Holdgate, Martin W., Kassas, Mohammed, and White, Gilbert F. (eds.), *The World Environment 1972–82*. Dublin, Tycooly International Publishing Ltd., 1982.

28. US National Academy of Sciences, *Science and Technology: A five-year outlook*. San Francisco and Oxford, W. H. Freeman, 1979.

29. Holdgate *et al.*, op. cit.

30. Wick, Gerald L., and Schmitt, Walter R. (eds.), *Harvesting Ocean Energy*. Paris, The Unesco Press, 1981.

31. Intergovernmental Oceanographic Commission, *Ocean Science for the Year 2000: Report of an expert consultation organized by SCOR/ ACMRR with the support of IOC and the Division of Marine Sciences of Unesco*. Paris, Unesco, 20 July 1982, IOC/INF-505, SC-82/WS/43.

32. US National Academy of Sciences, op. cit., p. 33.

33. Eckholm, Erik, *Down to Earth: Environment and human needs*. New York, W. W. Norton and Co., 1982.

34. Somewhat arguably, biological—as distinct from chemical— weapons have not yet been used in warfare.

35. Holdgate *et al.*, op. cit.

Chapter 7. Technologies

1. US National Academy of Sciences, *Science and Technology: A five year outlook*. San Francisco and Oxford, W. H. Freeman, 1979.
2. Oettinger, A. G., 'Information Resources: Knowledge and power in the 21st century'. *Science*, 209.4452, 4 July 1980, 191–8.
3. Danzin, André, 'Entreprise et évolution'. In *Connaissance politique*, No. 1, Feb. 1983, 116.
4. Oettinger, A. G., op. cit., 192.
5. A detailed account of the field, and its future prospects, was included in 'Science, Technology and Society: Interactions', a working paper produced for the 19–22 May 1981 meeting of the Unesco Advisory Panel on Science, Technology, and Society. Paris, Unesco, 1981, document SC/634/03.
6. US National Academy of Sciences, op. cit., p. 271.
7. See Clarke, Robin (ed.), *More than Enough: An optimistic assessment of world energy*, pp. 10–11. Paris, The Unesco Press, 1982.
8. Food and Agriculture Organization, *The State of Food and Agriculture 1977*. Rome, FAO, 1978.
9. Food and Agriculture Organization, 'Forestry for Local Community Development'. Rome, FAO, 1978, Forestry Paper 7.
10. Paper prepared for UN Conference on New and Renewable Sources of Energy, Nairobi, August 1981.
11. US National Academy of Sciences, op. cit., p. 325.

Part III. Choices

1. See, for example: Independent Commission on International Development Issues, *North–South: A programme for survival*. London, Pan Books, 1980. Organization for Economic Co-operation and Development, *Facing the Future: Mastering the probable and managing the unpredictable*. Paris, OECD, 1979.
2. Rahman, Ibrahim Helmi Abdel, 'Science and Technology: The development dilemma'. In *The Unesco Courier*, Nov. 1979, 4.
3. United Nations, *The United Nations Conference on Science and Technology for Development*. New York, United Nations, 1980.

Chapter 8. The scientific community

1. Ziman, John, *Public Knowledge: The social dimension of science*, p. 102. London, Cambridge University Press, 1968.
2. The Invisible College was actually the name given to the precursor of the Royal Society in London and was resurrected by Derek de Solla Price in his book *Science since Babylon* (New Haven, Yale University

Press, 1961).

3. Ziman, John, op. cit., p. 9.

4. Quoted in Rose, Hilary and Steven, *Science and Society*, p. 11. London, Allen Lane, Penguin Press, 1969.

5. Bacon, Francis, *New Atlantis*. In Bacon's 'Works', ed. Spedding, Ellis, and Heath, 14 vols., vol. 3, p. 165. London, 1857–74.

6. Ziman, John, op. cit., p. 140.

7. Quoted in Greenberg, Dan, *The Politics of Pure Science*, p. 12. New York, New American Library, 1967.

8. Price, Don K., *The Scientific Estate*, p. 278. Cambridge, Mass., Harvard University Press, 1965.

9. Holton, Gerald, 'The Double Service of Scientists and Engineers'. In Kinnon, Colette M., (ed.), *Scientists on Development: International scientific and technological co-operation today and tomorrow*. Paris, Unesco, May 1979, SC-79/WS/77(rev).

Chapter 9. The public and the media

1. *Report of the Royal Commission on Australian Government Administration (Coombs Report)*, vol. 1, sect. 10.7. Canberra, AGPS, 1976.

2. OECD, *Technology on Trial: Public participation in decision-making related to science and technology*. Paris, OECD, 1979,

3. Ibid., p. 45.

4. Nelkin, D., *Technological Decisions and Democracy*. London, Sage Publications, 1977.

5. OECD, op. cit., p. 53.

6. See, for example: Freedman, Lawrence, *Britain and Nuclear Weapons*. MacMillan, London, 1980. Thompson, E. P., and Smith, Dan (eds.), *Protest and Survive*. Harmondsworth, Penguin Books, 1980.

7. Cambridge Experimentation Review Board, 'Guidelines for the Use of Recombinant DNA Molecule Technology in the City of Cambridge'. *Report to the City Manager*, 5 Jan. 1977.

8. Ibid., p. 4.

9. The principle of 'innocent until proved guilty' is not, in fact, quite as obviously correct in this area as it is in general law. Several environmentalists, for example, have argued that all new commercial chemicals should be assumed to be environmentally hazardous until proved innocent.

10. Culliton, Barbara J., 'Science's Restive Public'. In Holton, Gerald, and Morison, Robert S. (eds.), *Limits of Scientific Inquiry*. p. 147. New York, Norton, 1979.

11. Nelkin, Dorothy, 'Threats and Promises: Negotiating the control of research'. Ibid., p. 191.

12. Green, Harold, 'The Boundaries of Scientific Freedom'. *Harvard University Newsletter on Science, Technology and Human Values* 20, June 1977, 17–21.

13. OECD, op. cit., p. 114.

Chapter 10. The nation state

1. Buzzati-Traverso, Adriano, *The Scientific Enterprise, Today and Tomorrow*, p. 390. Paris, Unesco, 1977.

2. *The Vienna Programme of Action on Science and Technology for Development*, p. 1. The United Nations Conference on Science and Technology for Development. New York, United Nations, 1979.

3. Rose, Hilary and Steven, *Science and Society*, p. 16. London, Allen Lane, Penguin Press, 1969.

4. Kapitza, P. L., 'Recollections of Lord Rutherford'. *Nature* 210, 782–3.

5. Ashby, Eric, *Technology and the Academics*, p. 94. London, Macmillan, 1958.

6. It is extraordinary how readily we have come to accept the use of this term, which reduces an educated person to the status of a commodity. According to Murray Bookchin, the process is intimately connected with the cause of environmentalism which 'tends to reduce nature to a storage bin of "natural resources" or "raw materials". Within this context, very little of a social nature is spared from the environmentalist's vocabulary: cities becomes "urban resources" and their inhabitants "human resources" . . . an issue more important than mere word play is at stake.' From *The Ecology of Freedom*, p. 21. Palo Alto, Calif., Cheshire Books, 1982.

7. The concept of grantsmanship is almost wholly due to the US science writer Dan Greenberg, who has immortalized the concept in a series of articles on 'Dr Grantswinger' in *Science* magazine. See also Greenberg, Dan, *The Politics of Pure Science*. New York, New American Library, 1967.

8. Barzun, Jacques, *Science, the Glorious Entertainment*. New York, Harper and Row, 1964.

9. Ravetz, J. R., *Scientific Knowledge and its Social Problems*. Oxford, Clarendon Press, 1971.

10. Gresford, Guy, 'The Responsibility of the Chemist in Development'. In *Proceedings of the International Symposium on University/Industry Interactions in Chemistry* (Toronto, Canada, 4–7 Dec. 1978). Paris, Unesco, 1979, p. 292, SC-79/WS/44.

11. *Science and Technology Statistics in Asia and the Pacific*, p. 3. Paris, Unesco, 1982, SC.82/CASTASIA II/ref 3.
12. *Science, Technology and Development in Asia and the Pacific*, p. 44. Paris, Unesco, 1982, SC.82/CASTASIA II/3.
13. Ibid., p. 9.
14. These figures are deduced from Figure 1 of *Science and Technology Statistics in Asia and the Pacific*, op. cit., p. 3.
15. ACAST, *World Plan of Action for the Application of Science and Technology to Development*, p. 55. New York, United Nations, 1971.
16. *Science, Technology and Development in Asia and the Pacific*, op. cit., p. 48.
17. Ibid., p. 47.
18. Barel, Jacques, and Clarke, Robin. Paris, Unesco, 1981, unpublished report.
19. WIPIS. Philadelphia, Institute of Scientific Information, 1978.
20. 'Methods for Priority Determination in Science and Technology'. Paris, Unesco, 1978, Science Policy Studies and Documents, No. 40.

Chapter 11. The transnational enterprise

1. Taquey, Charles H., 'The Transnational Enterprise'. In *Diogenes*, No. 105, Spring 1979.
2. Quoted in Sinha, Radha, 'Agribusiness: A nuisance in every respect'. In *Mazingira*, No. 3/4, 1977.
3. Dolman, A. J., 'The World Economic System and Third World Environment'. The Hague, The Foundation for Reshaping the International Order, Feb. 1982, mimeograph.
4. Clarke, Robin, and Palmer, Judi, *The Human Environment: Action or disaster?* Dublin, Tycooly International Publishing Ltd., 1983. The debate on environmental imperialism begins on page 52.
5. Quoted in Szekely, Francisco, 'Pollution for export'. In *Mazingira*, No. 3/4, 1977.
6. See Szekely, Francisco, op. cit.
7. Unesco, 'Meeting of experts on the orientation of Unesco studies concerning the influence of transnational corporations' activities in the Organization's fields of competence' (Helsinki, Finland, 12–14 Apr. 1978, Final Report). Paris, Unesco, 1979, SS-79/WS/6.
8. Mattis, Ann, 'Science and Technology for Self-Reliant Development'. In *IFDA Dossier 4*, Feb. 1979.
9. Hogbe-Nlend, Henri, 'Transfer of Technology'. In Kinnon, Colette M., *Scientists on Development: International scientific and technological co-operation today and tomorrow.*, Paris, Unesco, May 1979, SC-

79/WS/77(rev).

10. Abdel Rahman, I. H., contribution prepared for the Unesco Advisory Panel on Science, Technology, and Society. Paris, Unesco, 1981, unpublished.

11. Quoted in Brister, Judith, 'Development Issue Paper for the 1980s No. 7: obstacles to North–South technological co-operation'. New York, United Nations Development Programme, undated.

12. Dolman, A. J., op. cit., p. 29.

13. Schumacher, Fritz, *Small is Beautiful: A study of economics as if people mattered*. London, Blond and Briggs, 1973. For a brilliant rebuttal of the arguments of Schumacher and his followers, see Rybczynski, Witold, *Paper Heroes*. New York and Dorchester, Anchor Books and Prism Press, 1980.

14. Independent Commission on International Development Issues, *North–South: A programme for survival*. London, Pan Books, 1980.

15. Agarwal, Anil, Press Briefing Document on the United Nations Conference on Science and Technology for Development. London, Earthscan, 1979.

16. Unesco Advisory Panel on Science, Technology, and Society, 'Report of the Second Meeting'. Paris, Unesco, Sept. 1981, SC/631/07.

17. Ibid.

18. Unesco, *Bi-Annual Report on Priorities and Trends in Research related to Human Needs*. Paris, Unesco (in collaboration with the International Federation of Institutes for Advanced Study), Sept. 1978, SC-78/WS/54.

Chapter 12. The international dimension

1. Salam, Abdus, 'Science and Technology in Developing Countries'. In Kinnon, Collette M., *Scientists on Development: International scientific and technological co-operation today and tomorrow*. Paris, Unesco, May 1979, SC-79/WS/77(rev).

2. Salam, Abdus, 'The Candle of Learning'. In *The Unesco Courier*, Nov. 1979.

3. Ibid.

4. Ibid.

5. Ibid.

6. Unesco Advisory Panel on Science, Technology, and Society, 'Report of the Second Meeting'. Paris, Unesco, Sept. 1981, SC/631/07.

7. Unesco Advisory Panel on Science, Technology, and Society, op. cit.

8. Quoted in Salam, Abdus, op. cit.

Index